This book is dedicated in gratitude
and affection to Dorothy M. Bryan and
Raymond T. Bond, friends and editors

Acknowledgments

I am much indebted to Esther Hunn and other librarians at the Illinois State Library, and to Orvetta Robinson, Illinois State Museum librarian, for their success in securing for me certain rare books of reference which I might not otherwise have had the opportunity to see and use.

Grateful acknowledgments are due to Houghton, Mifflin Company, Boston, for permission to use quotations from John Muir's writings which originally appeared in *Life and Letters of John Muir*, by William Frederick Badé, 1924; *Cruise of the Corwin*, by John Muir, 1917; *John of the Mountains*, unpublished journals of John Muir, 1938; and *The Wilderness World of John Muir*, by Edwin Way Teale, 1954.

I am, as always, deeply indebted to my husband for his help in working with me and editing my manuscripts.

VIRGINIA S. EIFERT

Introduction

At least one common interest united such diverse men, things, places, and human experiences as Captain John Smith, Carl Linnaeus, the French Revolution, the Spanish Conspiracy, the Mississippi River, Lake Superior, Thomas Jefferson, Jefferson Davis, the Santa Fé Railroad, Henry Thoreau, and the discovery of Yellowstone Park. That single bond of unification was the great green landscape of America. It bound together people of the past with those of today, gave meaning to a continent's geography from ocean to ocean, and brought about the identification of its natural history dwelling from deserts to mountaintops.

Without that remarkable vegetation which fascinated so many persons, the American landscape itself would have been extraordinarily barren and uninteresting. Plants made history, and history itself would have been dull without the restless men who followed the call of the far horizon and ferreted out the intricate story of rivers, plains, mountains, and shores and, at the same time, discovered our plants and animals.

Natural history makes strange bedfellows—or rather, in this case, strange combinations of human personality, international crises, war, peace, discovery, and adventure—in terms of a nation's expansion and development. These trees and flowers of America, perhaps more than any other part of natural history—surely even more than its animal life or its geology—sparked an immense curiosity among Europeans. The incentive to discover new plants in America contributed importantly to a wider exploration of the entire continent.

Simply to blaze a trail from here to there was one thing, and it was a necessary thing; yet cutting a narrow route from one place to another left it still only a slender thread of knowledge drawn across an enormous landscape. But when the naturalists ranged those trails for specimens, combing broader areas on either side, a very much greater mass of discovery took place. This lay not alone among the wildlife but in the topography itself, in the locating of mountains and rivers, in the finding of new lakes and waterfalls, and in figuring the heights of mountain peaks. It lay in a knowledge of timber resources so important to a growing nation, in the medicinal and otherwise useful plants, in ferns and mosses, even in the fossil plants which, simply because of their antiquity and beauty, were all a part of the great and growing picture of America.

America, of course, would have survived without the men who made the painstaking and detailed discoveries and studies of trees and flowers. It didn't really matter to the welfare of the majority of the people if men were digging up and identifying hundreds of fossil plants from which to reconstruct knowledge of the extinct forests of North America. If no one had ever found the mysteriously rare Franklinia tree or ever had lost it again, the world and the State of Georgia would have been unaffected. If David Douglas had not risked his life to track down the western sugar pine, or if the German botanist Creutzfeldt had not found a certain small Gilia in Utah before the Paiutes killed him, or if Edwin James had not found the Colorado columbine on the slopes of Pikes Peak, then surely the pine would have been found by someone else, the Gilia identified later, and the columbine discovered by another climber.

But all of these things did take place, and they are all a part of America and of its history. This country and the world have been made very much the better, simply because knowledge is a priceless commodity and the search for it is an inescapable part of the inspiration and advancement of human life and its mental progress.

That search for knowledge of America's wild things began when the Norsemen, happening upon the coast of North America, came across wild grapes and wild rice and sugar maple trees, as well as certain edible roots, fruits, and seeds. It continued excitingly with the coming of Columbus to the islands and De Soto to the Mississippi, and with the arrival of Thomas Hariot and Captain John Smith in the Roanoke and Jamestown settlements of Virginia, at a time when starvation and Indians were surely of greater importance than the discovery of trees and flowers. Yet, even in the midst of trying and crucial times, discoveries were made and highly detailed reports went back to England. These challenged other men to add to the growing knowledge of a continent— knowledge which had only begun when the land itself was found. Unknown and unguessed in numbers and variety, everything inland from the coasts would have to be seen and named before America itself could be said to have been truly discovered or really known.

With the inevitable trend toward a westward expansion, plant hunters accompanied most exploring and pioneering groups; botanists took part or were slain in Indian attacks; they assumed vital roles in history and in the opening of the western lands. But no matter what the exigency or the occasion, these intrepid men were always essentially aware of their surroundings, were botanists and naturalists to the end. They might have worked alone, or may have gone as part of the well-outfitted government expeditions, but, however they went, and because of their single-minded efforts, America eventually became crisscrossed and trailblazed by so many naturalists, plant collectors, and scientists, from 1000 A.D. until and including the present, that thousands of plants have become known and the wildlife of the continent has been named and catalogued.

In North America alone, tens of thousands of plants ranging

from the algae and fungi through the ranks of lichens, liverworts, mosses, club mosses, ferns, Equisetums, coniferous trees, and deciduous plants—these latter varying from minute duckweeds to splendid Sequoias—all had to be found and named. One by one, year by year, they were discovered by plant hunters working for several centuries to accomplish their ends. Each newly found plant needed to be examined so that a careful description of all its parts could be written; it must show that it was unique in possessing this particular description. In order to make this possible, all the parts of the plant must be seen, measured, studied, the time of bloom determined, and the whole range of each plant from north to south and from east to west must be found out. All these details were not always accomplished by the original finder, but often by succeeding generations of botanists who, one by one, year by year, plant by plant, added new facets to the original discovery until all the facts were known. To multiply all this scientific labor by the tens of thousands of American plants reveals something of what has been quietly going on in field and forest and from mountaintop to desert ever since men with the eyes and minds of botanists first visited these shores.

As regions have been explored and herbaria collected, the whole tremendous knowledge of America as a continent has become understood in ways in which conquest and exploration alone would not have sufficed. The explorers might take care of the larger swath of discovery, of mapping the mountains, surveying the plains, and navigating the rivers and lakes, but it was the scientists who saw the infinite beauty of detail, and who brought it back to be added to the whole complex picture of our land.

Brought together in a flowering landscape, some of the men of discovery and some of the plants they found are represented in this book. They are a cross section of human history as it was inescapably combined with natural history, for nature and man can seldom be severed or unrelated to each other. It is the botanists, the plant men, who have known our tall trees and who have traveled toward our endless far horizons in search of the meanings of a green and growing world.

Contents

Photographs

The Catawba rhododendron was one of the discoveries of André Michaux and his son.

The flame azalea (Rhododendron calendulaceum) was one of the discoveries of Michaux.

The climbing fern (Lygodium palmatum) is still found along the valley of the Cumberland.

The dwarf lake iris (Iris lacustris) was discovered and named by Thomas Nuttall.

White marsh marigolds (Caltha leptosepala) blossom at the edges of old drifts on high mountains.

The wild hyacinth or camas (Camassia esculenta) blossoms in an expanse of pale lavender-blue.

Mariposa lilies were one of the charming finds of Nuttall, Douglas, and Bigelow.

The Mississippi River is a river of constant turbulence and change.

Spring came early along the Arkansas when Nuttall set out to discover new plants along its upper reaches.

This forest in the Porcupine Mountains of the upper peninsula of Michigan appears much as it did when the Cass expedition passed this way.

The bearberry (Arctostaphylos uva-ursi) was one of the plants found by David Bates Douglass.

Moss campion (Silene acaulis) grows in low mats on the rocks above timberline of high mountains and tundra.

Beyond the narrowing Mississippi lies Lake Itasca, headwaters of the Mississippi.

The Douglas firs populate whole forests.

At the edge of melting snow, David Douglas found small flowers emerging from the stern winter of the high country.

Even in the high mountains small flowers bloom in their brief summertime.

The Joshua tree was discovered by John Engelmann.

Yuccas of many species were seen on western expeditions.

The mesas of Utah, New Mexico, and Arizona were part of the landscape as expeditions worked their way to the coast.

The beauty of a fern is preserved in a nodule of stone.

Thoreau and his companions canoed on lakes and streams of the Allagash wilderness.

The purple-fringed orchis (Habenaria psychodes) was seen from the stage coach going to Moosehead Lake.

"It was a mossy swamp . . . ready to echo the growl of a bear, the howl of a wolf, or the scream of a panther. . . ." From Thoreau, *The Maine Woods.*

The twinflower has delighted naturalists from the time of Linnaeus to the present.

The fans of the palmettoes rise to meet the soft veils of Spanish moss hanging from live oaks and bald cypresses.

The Sequoias are the climax of American vegetation.

Sassafras

1. Land of the Sassafras Tree

When the ships, after much difficulty in the pounding breakers off Cape Hatteras—it was called Hatorask then, the Indian word—finally came into the calmer waters of Roanoke Sound and approached the silent shores and the empty stockade surrounding lifeless cabins on the island, the worst forebodings in the heart of John White welled to the surface. He was too late. Too late—they were all gone from there, and, even while he searched, he must have known that he would never again see his granddaughter, the little Virginia Dare, or any of the people to whom he had promised to bring help. He was too late. He had failed them, and now they were all vanished into the brooding silence of the American forests. One word, CROATOAN, he had found carved on a tree, but it was a cryptic message, and there was no other word, no explanation, no clue as to where the people had gone or what had been the fate of that colony. It was 1590, and John White had come back too late.

The tragedy had commenced only six years before, but it had not been tragedy then, only a project full of high expectation, good will, and the promise of success. England had been slow in getting a roothold in the New World. Spain for too long a time had ruled Florida, and, although its small settlements were feeble and the people enduring great hardship, the tales which reached England had been those of the wealth and wonder to be had here by any enterprising men or nations. In spite of sufferings, the people had spoken of the great prospects of the land, in proof of which they had sent back the bark and roots of a tree called sassafras. This was one of the truly great products used in medicines and cure-alls, and it was only one of the items of value in America. The land was obviously too full of such profitable natural products for England to be able to stay out of the race for dominance of the New World. Queen Elizabeth in 1584 thus gave Walter Raleigh letters patent to set up a colony in Virginia. If an English colony could be maintained for at least six years, it could be assumed that England had in fact struck roots and might lawfully claim the land for the queen.

Raleigh was a cautious man. Before sending any colonists, he dispatched Philip Amadas and Ralph Barlowe, in two ships, to discover what lay along the American coast and to choose a place which they felt could sustain a colony. While they were there, Amadas and Barlowe were to look for sassafras. Its presence might in fact influence the location chosen for the colony itself.

Captain Barlowe was enthusiastic about what they found. He and Amadas, besides, were in Virginia for only six weeks, which was not long enough for them to have obtained any very unfavorable impression of the inhabitants or of the land itself. Barlowe wrote:

On the second of July we came into shallow water, and the air smelt as sweet and strong as if we were in a fragrant flower garden. . . . The grapes grew everywhere, on the sand, on the green soil of the hills, over the plains, climbing on every little shrub and toward the tops of the high cedars. . . . The woods are not barren and fruitless . . . but are thick with the highest and reddest cedars in the world, far better than the cedars of the Azores, of the Indies, or of Lebanon. Pine, cypress, sassafras, and the lentisk, or gum tree, all grow there, as well as a tree whose bark is black cinnamon, which Master Winter brought back from the Straits of Magellan; and there are many other trees of excellent quality and fine fragrance. . . . We saw at least fourteen different sweet-smelling timber trees and the greater part of the underbrush is bay and growths of that nature. They have the same oaks that we have in England, but far larger and better.

Ralph Barlowe and Philip Amadas scouted along the offshore islands and capes of the Carolinas and Virginia, found inlets lead-

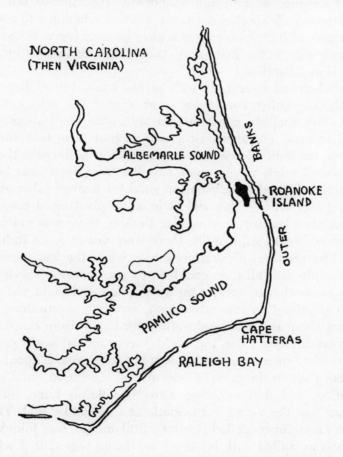

ing through Cape Hatteras (or Hatorask), into Roanoke Sound and the lovely Roanoke Island. They felt that this would be a perfect location for Raleigh's colony. It was happily sheltered from the storms and comfortably adjacent to the mainland. The

Indians seemed tractable; the two captains had been able to maintain fairly good relations with the inhabitants.

Next year, therefore, on the strength of this report, five small vessels—the *Lyon,* the *Tyger,* the *Roebucke,* the *Elizabeth,* and the *Dorothy*—sailed out of Plymouth harbor on an April day in 1585, bound for the fair and promising land of Virginia. They carried supplies which were confidently expected to last for a year, after which Sir Richard Grenville was scheduled to come at Eastertime with replenishments and news from home. Since there were only men in the first group, they would be hungry for word of wives and families.

The whole affair seemed well planned and full of hope, yet, from the beginning, everything went wrong. The colony did not thrive. The English, constantly at war with the Indians, committed so many outrages on the natives that they, in natural retaliation, made life precarious for the colonists. Because the men were almost solely relying upon supplies from home, and besides considered themselves a little too good for manual labor and for plowing and sowing, they did little or no planting of crops that summer after landing on Roanoke. Besides, there was much sickness; men died of strange ills; there were deaths from Indian attacks. The numbers of men grew less, while the leaders, setting the example, had all they could do to persuade the survivors to get to work—that to survive meant to plant and build and think, not to sit about in discontentment, swatting mosquitoes, complaining about the food, and waiting for succor from England. At this great distance across the world, succor indeed might prove to be very slow in coming, and it was not to be relied upon.

There were in the group a few men of resolution, intelligence, education, and determination. Governor Ralph Lane was one. Another was the young mathematician and intellectual, Thomas Hariot (sometimes spelled Harriot). Still another was John White, who was as skilled with his brush as Hariot was skilled with his pen. Between them—Hariot and White—it is possible to know what it was really like in Virginia in 1585, for the one wrote an extensive, detailed account of Indians, wildlife, and landscape, while the other made the first water-color paintings and sketches of wild animals, birds, fishes, trees, and flowers in America. Long after the colony had failed and its failure had been almost forgotten, the reports of Hariot and White remained. They are living pictures in paint and words of what the American wildlife was all about, there along the outer fringes of the continent where the storms of Hatorask battered the sands, and the trees, as Barlowe had said in the beginning, were fragrant and huge and the most beautiful in the world.

John White was in America thirty-five years before the Plymouth colony was begun in Massachusetts, and ninety-three years after Columbus had made the landfall which marked a turning point both in the world's history and in the development of natural history. Columbus had changed a flat world into a round one. John White, Thomas Hariot, and the naturalists who followed them were to turn an unknown wilderness into a place of knowledge, wonder, and inspiration.

The early naturalists in America were hampered by their own lack of knowledge, by their inability to find solutions to their problems, and the impossibility of getting answers to their questions. Instead of relying upon men of science in Europe to solve the problems, they themselves were going to have to work out the solutions and the answers. From the start, it was seen that the North American flora and fauna were obviously unlike those in Europe and the Old World, although there were indeed some similarities and some likenesses which only made matters more confusing. It would take a good many years, several centuries, in fact, before more than a dent was made in discovering and identifying the hordes of American plants and animals which were largely so different—different, exciting, and, for a long time, incomprehensible—from those of the Old World.

The only guidebooks in those days were the herbals of doctors and the manuals of the alchemists. They were written laboriously in Latin or Greek and were illustrated with woodcuts of often doubtful realism, although they were beautiful in design and detail. But in America these books had little relevance. There was no way in which to identify the specimens discovered except to find varieties which were somewhat similar to those in Europe or, when no likeness at all could be found, simply in desperation to invent names for them.

It must have been a frustrating situation for the men who really wanted to know and who had no place to turn for information because there was no information to be given. As pioneers and pathfinders, trailblazers in the wilderness not only of America but of the mind, it was these men who set the foundations for the later slow, orderly development of natural science in America. The great gathering of knowledge was a long time in coming, but the wonder was that it came as soon as it did. When a man had to be more immediately concerned with whether or not he might be murdered by Indians in the night, or if he would find enough food to feed himself and his family for another day, or another year, he could not be too much concerned with ferreting out the names of flowers and trees which neither protected nor fed him in the fearful forests where savages and starvation lurked.

Nevertheless, there have always been men who, in spite of everything, could not resist the proddings of curiosity. Despite hunger, disaster, and death, they still, as White and Hariot did, had to know something about their surroundings.

They belonged to an era in which scientific inquiry was groping out of the Middle Ages into the realm of discovery. Hariot, later to become a noted mathematician and astronomer, and the better educated of the two, in making his report, was following the custom of others before and after him in featuring the uses of the local products. He spoke of the pines because they gave pitch and tar, resin and turpentine, and he could not fail to extol the virtues of the wondrous sassafras which was usually the first thing an explorer looked for along the American coasts—"a wood of the most pleasant and sweet smell and of rare virtues in medicine for the cure of many diseases."

Hariot knew the cedar, the walnut, and three kinds of oak—oaks whose acorns, he decided, might be used to make an excellent oil. He mentioned plants which offered dyes. There were the Indian foods, especially maize, a most wonderful item of diet, a salvation to the starving; and beans, peas, pumpkins, melons, and squash, as well as a plant called orach which, he said, was "used as spinach, an excellent pot-herb." He found sunflowers, called *planta solis*, from whose seeds the Indians made both bread and broth.

He experimented with the native tobacco. Sir Walter Raleigh is credited with having introduced it to England, and in a measure he did, but it was Thomas Hariot who is said to have brought it to him. Hariot wrote: "The fumes . . . not only preserve the body, but if there are any obstructions it breaks them up. By this means the natives keep in excellent health, without any of the grievous diseases which often afflict us in England."

Hariot walked along the shores, penetrated the forests to the Indian towns, explored the edges of marshes, and ventured to the hills lying beyond. He must have found wild rice—"There is also a variety of reed which bears a seed much like our rye or wheat and when boiled, makes a good food." He found wild crab apples, prickly pear cactus, wild strawberries, mulberries, and whortle-berries; mentioned such trees as the cottonwood, maple, witch hazel, holly, willow, beech, ash, elm, and a tree he identified by its Indian name, *Ascopo*, because he could compare it with no English tree. It may have been the prickly ash with its spicy, burn-ing, inner bark and tiny, peppery berries. The Indians used them as a counterirritant in a bad case of toothache. Packed into a hollow tooth, the fury of the little berries could be expected to make the sufferer forget his pain in the fire which suffused his mouth and then caused a blessed numbing effect. *Ascopo*, to men far from dental aid, could be a boon.

In spite of all the things he did recognize or could describe, Thomas Hariot concluded his detailed account of the vegetation of Virginia by sadly recognizing the fact that he had really seen very little in America. He had explored only in a brief area, had discov-ered hardly any of the undoubted botanical riches. "Many other strange trees can be found here, whose names I know only in the Virginian language," he wrote.

Somehow, in spite of the hardship and hindrances, he and White managed to record their impressions of wildlife in America. How they did so remains a marvel, for matters in the colony were in-deed going poorly, and, as time passed, seemed disinclined to

improve. As the winter came on—a lovely, mild, Virginian inter-
lude, scarcely recognizable as winter—a deadly apathy lay over
the men of Roanoke.

More than half of the original number had died. Daily now the
survivors, looking for a sail beyond the sand spits of Hatorask,
watched the eastern sea, but they saw only the laughing gulls and
the red-beaked skimmers, and the big green-gray waves piling in.
The governor almost had to force the men to bestir themselves
and to plant gardens. The supply ships, he knew, might truly
never come. The watchers, however, continued their vigil.

Time went, and it was spring. It was Easter, and still no ships
had come. In the face of new Indian trouble, the men at last
planted gardens as an emergency supply of food, and lethargically
hoed the sprouting beans and corn. The vegetables were growing
well when, suddenly on a hot day in June, a great shout went up
from the sea-watchers, while those who were working dropped
their axes or hoes, and the sick raised their heads from their fetid
pillows. Twenty-three ships lay in the sea-glitter out beyond the
inlet. The Roanoke men had prayed for even one, but God had
sent them a whole fleet.

It was Sir Francis Drake with his armada on his way up the
coast from plundering the Spanish forts in Florida. He sent men
ashore to see if they could do anything for the Englishmen on
Roanoke. They were met with a clamorous eagerness to be taken
home—the colony was failing, it was desperate, the Indians were
menacing, white men could not survive here under such hard
conditions. Besides, they were homesick for England. Therefore,
in spite of the urging of men like Lane and White and Hariot, who
felt that some men at least should remain to hold the colony for
England, every man hastily packed up his belongings, and, after
some difficulty, managed to get aboard the ships. John White took
his portfolio of paintings. Hariot took his journal and specimens
of plants for Raleigh. Roanoke, with its stockaded fort and the
huts inside it, and the gardens where corn and melons and beans
were growing, was left to the Indians, the deer, and the birds.

Two weeks later the longed-for supply ship of Sir Richard
Grenville, delayed by trouble with the Spanish, reached Roanoke.

Sir Richard found no one left, and there was no indication of where everyone had gone. Seeking some trace of the colonists, Grenville ranged inland, along the coast, and among the islands, and then he gave up. He decided they had all been killed by the Indians. However, not wanting to leave the colony unclaimed, he delegated fifteen brave and sturdy men to hold the fort while he sailed back to England with the information. He deposited with them enough food and clothing to last for two years, but promised that ships would surely be back the next year to bring more colonists, together with farm animals, equipment, and tools.

When the first Roanoke colony had fled with Drake, Thomas Hariot had gone against his will. When he wrote his detailed report of the whole experience in America, he expressed some of his love and admiration for the place and its possessions, and at the same time uttered a good deal of resentment and criticism for the laziness of the colonists who had caused it to fail. He directed his report to the farmers, investors, and well-wishers of any further attempt to colonize and plant in America.

For Hariot, the longer he thought of what had happened, was downright indignant. He was tired of hearing the new land slandered and made little of, for he realized that the failure in colonizing, that first time, had been caused by the people themselves, not by any lack to be found in America itself. The new country was good, but the people—they could certainly have been improved. He wrote scornfully:

Some of them were ignorant of the country. When they returned to their friends and acquaintances they pretended to know more than other men, especially when there was no one in the gathering who could disprove them. They made themselves out to have suffered greater hardships than anyone ever suffered before. They put such value on their reputations that they would have thought themselves disgraced if, after living in Virginia a year, they had not had a great deal to say, true or false. Some of them spun tales of things they never saw. Others shamefully denied happenings which they did not see, but which nevertheless were known to have occurred. Still others made difficulties of simple things because they did not understand them.

The cause of their ignorance was mainly that they never left the island where we had our settlement, at least not to go far from it, during the whole time we were there, and therefore had seen little. Or they had lost interest when they did not immediately find gold and silver and spent their time in pampering their bellies. And there were others who had not much understanding, less discretion, and more tongue than was necessary.

Some of them had been nicely brought up, living in cities or towns, and had never seen the world before. Because they could not find in Virginia any English cities, or fine houses, or their accustomed dainty food, or any soft beds of down or feathers the country was to them miserable, and they reported accordingly.

On the strength of Thomas Hariot's arguments and the insistence of some people to go out to Virginia to make their homes, there were one hundred and seven persons, including a few women and children this time, who set out in 1587 for Virginia. John White was going back. He had been designated governor of the new colony, and with him were his daughter and her husband, Elinor and Ananias Dare. Elinor's first child was to be born very soon after they had arrived in America.

It was neither a good nor an encouraging trip. There were many disappointments along the way, especially when Captain Fernando refused to stop in the West Indies to get supplies of salt, sheep, cattle, and plants of pineapple, banana, and other fruits which were to be experimented with in the colony. The ships came close to being wrecked by this same Fernando's stupidity, and it was considered to have been a dispensation from heaven that they were permitted at last to arrive on the shores of Hatorask, that long, narrow, sandy cape east of Roanoke Sound and Roanoke Island itself.

They arrived in July. When they had rowed across to a silent and somehow frightening Roanoke Island, they found no one there. The fifteen men who had been left the year before by Grenville were gone. There were only some human bones as a reminder of what might befall every one of them. The fort built by the first colony was largely ruined, but some of the houses were still standing, and so was most of the stockade. The fort area was overgrown with wild melons, no doubt seeded from the gardens which the first colonizers had left behind. At the newcomers' approach, half a dozen deer jumped away from feeding on the melons.

There was no time to give way to fear and discouragement. Since John White had been there before, he had the advantage of knowing something of the lay of the land, the natural resources, and about which Indians were friendly. Orders were issued. Every man was to get to work repairing the houses which still stood and building as many more new ones as should be needed. It was suddenly very good to be on land, to be at work, to feel the warmth of the Virginia sunshine, and to breathe the fragrant air. No Indians made an appearance. The island and the mainland seemed to be deserted. The colonists hoped that the Indians might not discover they were there, would allow them the peace in which to build and repair and to erect a bulwark against any subsequent trouble.

But five days later when George Howe, in an effort to catch some crabs for dinner, was wading in the shallows, he fell with an arrow through his throat. No one saw his assailant. They only saw him fall face forward in the water, and when they ran to see what had happened, they found him dead and his blood staining the restless sea water. After that, nothing on Roanoke was quite the same. Fear replaced peace.

Nevertheless, the people worked. They planned for the future. Governor White was pleased with their industry. He was even more delighted and relieved when his beloved daughter Elinor gave birth to a girl-child on August eighteenth. They named her Virginia because she was the first English child to be born in the new land. Life improved behind the high stockade with the growing gardens.

Still, matters were not quite right. A few Indians were friendly, but there was always an undercurrent of menace. Men feared to go far afield on the mainland. At the same time, they saw that the supplies which had been brought from England were not going to be adequate to maintain the people until the next ships came in the following summer. There were meetings at the fort, and it was agreed (by all but White himself) that he, as the most influential man of the colony, must go back to England when the ships returned, plead for more supplies, especially livestock and arms, and see that they were sent quickly. White did not in the least want to leave Roanoke. He had just become a grandfather, and he wished nothing more than to stay here with his family.

But the people were urgent, and they were frightened. They needed in England a representative to speak on their behalf.

With a terrible sense of foreboding, John White kissed his daughter and the baby good-by and set sail when the ships departed in late August. It was a frightful voyage of many storms and heavy buffetings. The sailors felt they would never see England again. When the ships at last reached Plymouth, John White hurried at once to talk to Raleigh about the pressing needs of the planters in Virginia. Raleigh regretted the situation, was furious that Captain Fernando had not secured the necessary livestock and other items in the islands. He gave White money to outfit ships and buy all the supplies he needed, so that he could sail as soon as possible. Even in the face of the storms of winter, he would go.

But the more speed John White urged, the more everything dragged. The timing was all wrong. Men, women, and children, including a very young baby, might starve and die in the wilderness, but the more immediate needs of England must first be served. In the war with Spain, Philip's armada was just then poised in awesome might to attack English ships. Queen Elizabeth, in this crisis, prohibited all but warships from leaving the harbors.

Across the Atlantic, on a small, sandy, pine-shadowed island, there were crises, too, but no one in England knew about them. None but the desperate John White seemed to be concerned.

Continuing to hammer at the emergency, he finally managed to secure two small ships and the supplies, and hired fifteen men to handle the vessels. In April, 1588, they departed secretly. But he should have been more suspicious of his sudden ability to procure ships and men. The sailors and captain had used him. Once they were at sea, instead of sailing for Virginia, they made it their business to search out Spanish ships as prizes, to be attacked and taken at great risk. It was the most popular sport of the day.

The two vessels, however, were too small and poorly defended for this dangerous fun, and in an encounter with a lusty Spanish ship they were both nearly sunk. The two battered little craft, with the anguished John White still aboard with his futile supplies for Virginia, turned back to England.

Yet, even that summer when the Spanish Armada was defeated and English seamen could sail in safety again, no relief was sent out to America. Raleigh was in financial straits. The years 1588 and 1589 passed without any way in which John White might get himself back across the ocean to where he most wished to be, to the place about which he now had the most dreadful premonitions of disaster. England had deserted the Virginia colony.

Not until 1590 did John White finally find passage on one of three ships bound for America. He was refused permission to take anything but his own sea chest—no supplies, nothing for the stricken people. The ships, again ostensibly heading straight for Roanoke, again turned buccaneer, as was the fashion of the times, to chase down Spanish vessels and rob them. Another tedious and anguished summer passed before the impatient and frantic John White finally saw the sandy shores of Virginia lying beyond the crashing breakers of Hatorask.

It had been three long, desperate years since he had left those shores. As the ships anchored in the heavy seas off the cape, he saw with a surge of thankfulness that, over on the island, a column of smoke arose. He was fairly flooded with relief at the sight. But when they headed to the land next day, battling the rising surf of a storm which was building in the Atlantic, perhaps the advance of a hurricane, the small boats starting out were almost swamped. Two of the sailors were washed away and drowned, and the rest were so alarmed that they wanted no more of this place. They wanted only to go home.

Smoke still spiraled faintly from a deserted campfire when at last the men landed on Roanoke Island, but no people were in sight. The newcomers shouted. Thinking the inhabitants feared that the ships were Spanish, they ordered their trumpeter to play all the dear, nostalgic English songs he knew, hoping that the familiar music would bring the colonists from hiding. The fort, however, was vacant, hollow, ruined, ravaged. On a tree were the letters CRO, carved there rudely with a knife. On another tree was CROATOAN. Croatan was the name of an Indian town fifty miles away, where they had long ago talked of going. John White remembered.

He had worked out a prearranged sign with his people before he left, so that if he came back and the village had indeed been moved, he would know where to find them. They must carve a message on a tree, he had said, and if they were in trouble, they were to place a cross above the letters. But there was no cross, and the message, if such it was, was a cryptic one. John White and the others hunted as long as they dared for a clue as to what had happened, but the rising tempest was causing deep apprehension among the men. The breakers were thundering off Hatorask and the wind, lashing the pines and cypresses, was tearing off twigs and sending leaves flying. The captain feared for his anchor cables, and by morning, in fact, a cable indeed had snapped and the anchor was lost. The ships were riding wildly on the great waves rolling toward the treacherous reefs and shores of the cape. There was no way for anyone just then to travel fifty miles to the Indian town of Croatan or to search elsewhere for the lost colony.

Instead, much against White's pleas, the captain ordered everyone aboard. Promising White to return in the spring with supplies, he intended going to Porto Rico for the winter. Again the weather defeated their plans. They finally had to go back to England; were, in fact, almost blown there by the hurricane.

There was no knowledge of what had happened on Roanoke. There never has been any solution to the mystery. The people of Roanoke, including the little Virginia Dare, simply disappeared, and with them went their history. One hundred and seven people vanished. There was only the grieving John White, and there were his records and his paintings, which tell something imperishable, revealing the pleasanter aspect of what Virginia was like when a naturalist looked for the first time at the wildlife of America.

John White was only the first of many who were to come. For a long time, the naturalist in America would by necessity be governed by need itself. In particular, one vital tree was sought, and the explorers and adventurers, though they were really not botanists, very quickly learned to identify it. Sassafras was not only distinctive in leaf shape, with its three forms of simple and lobed, aromatic leaves, but it was beautiful in autumn with its orange, scarlet, and yellow hues. Moreover, it had the sign of wealth imprinted all over it, and this naturally caused it to stand forth among less profitable vegetation. If a man in America needed to be able to recognize only one tree, it was this—the sassafras.

Sassafras was what Barlowe and Amadas had looked for. It had been what the Spanish, French, and English, wherever they touched on the American coast, first sought. After the fiascoes at Roanoke, the English continued to send ships to America—not to colonize, but to bring back cargoes of sassafras. Prices for this commodity soared rapidly in European markets. It was worth a man's while to get into the business. Thus in 1603, Martin Pring came on a commercial venture for this purpose. Landing in Maine, too far northward to find the trees he wanted, he determinedly worked his way south down the coast until he came to woodlands where the attractive, spicy trees grew abundantly. He taught his crew how to recognize sassafras, then sent them out into the woods to cut and dig. He and his men were so successful in identifying the sassafras in all the great confusion of the woodlands of a pristine landscape that he sailed back with two highly marketable, fragrant shiploads of the roots and bark.

Although the Indians had known sassafras chiefly as a blood purifier and tonic, and in the South customarily added the powdered dried leaves as a thickener to soups and gumbos, Europeans found a great many real or fancied uses for it. Sassafras became so highly thought of as a remedy for venereal diseases that, after a while, English gentlemen were embarrassed to be seen drinking the tea in public lest someone should think they were infected and were taking the cure. Sassafras had many other uses—to prevent bedbugs and moths, to flavor soaps and beer, to treat fevers, broken bones, and falling hair, to increase vigor and improve the

general health. It had much the same exciting impact upon six-
teenth- and seventeenth-century Europe as penicillin, sulfa drugs,
and the Salk vaccine have had upon civilization in this century.

A hope for a monopoly on the sassafras trade may have been
the chief reason why a third English colony was attempted in
Virginia. This time it was set in a different location. Cruising along
the island-dotted coast of Chesapeake Bay, the ships moved up the
inviting waters of the James River. Although the fact was not
emphasized, Jamestown was placed—certainly not by mere chance
—in some of the best sassafras country in Virginia. Here it grew
in dense thickets and whole woodlands, sending up suckers from
roots and quickly reforesting areas soon after they were cut.

When the ships went back to England, they carried from Jamestown a cargo of sassafras. This effort was largely due to the energies of a certain Captain John Smith, an enterprising man with
the alert mind of a speculator and the soul of an adventurer. He
realized that the best thing those ships could take back to impress
the stay-at-home English was sassafras—lots of sassafras. It had
not been easy, however, to persuade the new colonists to labor so
hard in digging, cutting, and preparing the bark for shipment, but
Smith insisted. He could be a martinet when necessary, and he
was becoming irked with the laziness around him. It was like the
first Roanoke colony all over again. Few cared to work.

And, as at Roanoke, when the loaded ships had gone again and
hunger pervaded the town, matters grew desperate. George Percy,
a naturalist and chronicler of the colony, said, "some departed
suddenly, but for the most part they died of meere famine. There
never were Englishmen left in a forreigne Countrey in such mis-
erie as wee were in this new discovered Virginia." Then, when
times were most desperate, the Indians, instead of destroying
them, came with mercy and with corn to save them.

The colony could have used more men like tough Captain John
Smith, a campaigner and veteran of the Turkish wars, who had
knocked about the world most of his life and who could cope
with almost any situation which presented itself. He was not the
sort to lie about in Jamestown and complain, grow hungrier, and
perish of anything less than an Indian's arrow. He was, also, an

excellent and knowledgeable observer of the settlement's surroundings. Like Hariot and White, he realized that the salvation
of the English colonies lay in their ability to use available resources
rather than in trying to live on the dubious and often long-delayed
supplies and promises from England. He wrote:

. . . all the Country is overgrowne with trees . . . many of their Okes
are so tall and straight, that they will beare two foote and a half square
of good timber for 20 yards long. . . . In some parts were found some
Chestnuts whose wild fruit equals the best in France, Spaine, Germany,
or Italy, to their taste that had tasted them all. . . . Plumbs there are
of 3 sorts. The red and white are like our hedge plumbs; but the other,
which they call *Putchemins* [persimmon], grow as high as a Palmeta.
The fruit is like a medler; it is first greene, then yellow, and red when
it is ripe; if it be not ripe it will drawe a mans mouth awrie with much
torment; but when it is ripe, it is as delicious as an Apricock.

Persimmons and chestnuts were indeed excellent finds, but he
knew that when he had found the sassafras he had struck an
arboreal gold mine. With this he could find an immediate market
and thus from the start would make the colony self-supporting—
that is, if he could persuade the men of Jamestown to get out in
the woods and dig. They were, to John Smith's increasing wrath,
seldom so inclined. Eventually, therefore, and in short order, the
Jamestown colony failed.

Yet, in spite of local failures, the knowledge of America's most
famous tree, from Florida to Canada, was spreading. Father
Claude Dablon, superior of the Canadian diocese in Québec, about
1680, in writing his annual report to Jesuit headquarters in France,
said:

But the most common and most wonderful plant in those countries
is that which we call the universal plant, because its leaves, when
powdered, heal in a short time wounds of all kinds; these leaves which
are as broad as one's hand, have the shape of a lily as depicted in
heraldry; and its roots have the smell of the laurel. The most vivid
scarlet, the brightest green, the most natural yellow and orange of
Europe pale before the various colors that our savages procure from
its roots.

After sassafras, certain newly discovered American plants found immediate commercial markets—the white pine, live oak, and maple timber were important from the first, and so were maidenhair ferns, from which the Indians had procured a valuable medicine for the relief of diseases of women and for ease in childbirth. The French thereupon turned maidenhair ferns into a profitable trade. But one of the truly vital discoveries among American plants, one which ranked next to the sassafras tree itself, was the ginseng, *Panax quinquefolius*.

Ginseng

When Father Joseph François Lafitau, a Jesuit, discovered ginseng in Canada, he had come upon a botanical bombshell. In the Orient it was a cherished drug to which were attributed wonderful restorative and curative powers. Better still, although it grew in China, the Chinese were providentially in short supply and would pay a great deal of money for ginseng in quantity. It was suggested to the Jesuits in Canada that they search for this plant in the woods around their missions.

Lafitau in Québec was enough of a botanist to know that some Chinese and American plants were very similar. Hoping to be successful in his hunt, he set out to search for the elusive ginseng, taking as much time as he could from his work with the Indians, and downright neglecting parish and parishioners. Looking for a plant that stood fifteen to twenty inches high, had two stalks with five to seven compound, palmate leaves held horizontally at the top of each stalk, and had, rising between the leaf stems, a small ball of greenish flowers which were followed by flat, scarlet fruits in a cluster, he covered miles of wilderness woods.

The Jesuit found a great deal of the wild sarsaparilla, which was in the same family and greatly resembled the ginseng, but it did not have the manlike shape of the stout, bifid root. He found the small dwarf ginseng, but this was not the right thing either. He had almost given up hope when, by chance in 1715, he discovered the true ginseng growing near a Mohawk Indian dwelling. The alerted Indians were set to digging as much as could be found. Ovens were set up nearby for the smoking and curing process. A major business had materialized in the Canadian wilderness.

By 1717, news of the bonanza had reached as far south as Green Bay, Wisconsin, where the Fox Indians set about securing ginseng in quantity, sending it to Canada to be cured, smoked, and exported to France, and thence, at a tremendous markup in prices, to the markets in Hong Kong. Soon the prices soared. There was much speculation in the new commodity until, through greed, the trade was almost ruined—for men had grown too hasty. A large shipment of ginseng roots, dug at the wrong season and improperly cured, had spoiled en route but had been sold anyway. Since China had no wish to purchase inferior ginseng, the market dropped. Although the trade with America was not totally concluded, it was, after the crash in 1754, only small and relatively unimportant.

From sassafras in Virginia to ginseng in Canada, white men for their own purposes and profit were beginning to learn of the plants of America. As more and more reports went back to England from men like Hariot and Smith, other men found their curiosities whetted, not only by the lure of commercial plants, but by the oddities, the botanical souvenirs brought back from a new and extraordinary land. Long before 1600, in fact, the Indian corn and the pitcher plant, as well as columbine, milkweed, everlasting, and arbor vitae, were known to botanists and herbalists in Europe. The pitcher plant itself must have been one of the true curiosities of the New World. It had no known uses, but it was so strange in appearance that it at once became a standard curio for travelers to bring back from America. Although a southern species, the

Sarracenia flava, may have been sent directly by Hariot from Virginia, or, before that, by a Franciscan monk in Florida, the northern pitcher plant came by a more roundabout way to the attention of the botanists. Carolus Clusius, a noted French scientist of the early seventeenth century, was said to have obtained a specimen from a Parisian apothecary who had received it, he said, from someone in London, who had purchased this oddity from a sailor just in from Newfoundland, where he reportedly had collected it. However, the specimen which was first named was sent to Paris

Pitcher Plant

from Canada by Dr. Michel Sarrasin de l'Étang, a physician at the court of Québec, and named *Sarracenia* in his honor. In Canada the plant was variously called *Herbe Crapaud, Petits Cochons,* or Indian jugs.

The evening primrose was known in Europe as early as 1614, and the herbal of John Gerarde not only contained this but pictures and descriptions of at least a dozen species of plants which had been sent from New England as curiosities to Mr. John Tradescant of London.

In 1635, the botanist Cornuti presented more American plants in a European book than anyone ever had seen before. His history of Canadian and other new plants of America was published in Paris with copperplate engravings of thirty-seven American species which from various sources had come to him. Men were still groping. None of them, thus far, had access to any representative collection from any one region in America. They simply used the oddments of plant curiosities which, by chance and good fortune, came to their desks, their laboratories, or their gardens.

When, in the eighteenth century, serious men of science—Mark Catesby, John Mitchell, John Clayton, and Cadwallader Colden among them—botanized extensively in America, it was they who truly commenced the great task of identifying American plants—as plants, not as curiosities or commercially exploited species. Yet, even before these indefatigable men were born, and at a time in history when American botany was still a resounding void of ignorance, New England had been visited by a single rare botanist named John Josselyn. He was, without doubt, the first of his kind in America.

Josselyn, an English gentleman, made two voyages to our shores. First in 1638, and again in 1663, he came ostensibly to visit his brother in Maine, although, that first time, it had been whispered about that John Josselyn had been forced to leave England because of personal political troubles. He was something of a scholar, well read for his day, and he possessed a considerable knowledge of science which was unusual at that time. Fascinated with what he found in America, he eventually wrote and illustrated with charming, amateurish drawings a book called *New-Englands Rarities Discovered*.

He had been so pleased with everything, and that pleasure is reflected in his book. Here were the fantastic New World marvels—the hummingbird, the skunk, the wild cat (which he called the ounce), and the groose, or grouse. He gave recipes and recommended remedies for aches, shrunken sinews, and wind on the stomach. His book included a goodly list of the fishes, but named only three insects—wasps, fireflies, and a large glittering beetle, perhaps the tiger beetle, which smelled odiously when crushed.

When Josselyn reached the category of the plants, however, it was plain that as a botanist he enjoyed them the most, looked at them with intelligence, and, moreover, knew a good deal about them. He had a copy of Gerarde's *Herbal* by which to identify his finds, but, since Gerarde listed and illustrated only a comparatively few American plants, Josselyn was confronted with immense quantities of species which were quite unknown and were impossible to identify. He described some of the unknowns so well, however, that botanists two centuries later could know what he had seen.

Unorthodox in naming and unscientific in wordage as Master Josselyn might have been, he did find, identify in his own way, and describe more than two hundred species of plants growing in New England. No man in America before him had troubled himself to find and recognize so many or had bothered to try to name many of those which he did find. There would be many

From Josselyn's "Hollow-leaved Lavender" or Pitcher Plant

more botanists who would come to collect and catalogue, but perhaps none of them would botanize with both the pioneering vim and the thrill of discovery, coupled with an astonishing candor and a vast lack of knowledge in the subject, with which Josselyn did in 1638 and 1663. Besides, after this, no matter how many more species the later botanists would discover—Cadwallader and Jane Colden, only a little less than a century later, managed to more than double Josselyn's list in New York State alone—none would work in quite the same situation as the fortunate Josselyn. They would all be hunting in a territory which was no longer pristine and totally untouched for, long before they had been born, John Josselyn, first in the field, had been pottering about in the woods and along the shores and marshes of New England and Maine, collecting specimens and giving to them the names which were first to be attached to some of America's trees and flowers.

Basswood or Linden

2. Linnaeus—a Name for Naming

Where the spruce trees grew smaller and more stunted, where the misty sweep of the tundra began, and where botanical discovery beckoned, young Carl Linnaeus on a day in 1730 began his big adventure. He, the son of an impoverished country pastor in a backwoods district of upper Sweden, had been sent to Lapland on one of the first—if not the first—botanical collecting expeditions. His way had been financed by the august Royal Scientific Society and by the king. Best of all, and unusual for his time, he was not required to look only for useful plants, but might collect everything he found in order to piece together a record of exactly what Lapland was like during its summer of virtually endless sunlight. To a young man in love with adventure, with travel, and most of all in love with botany, this was the doorway to paradise. Linnaeus was twenty-three years old and was just beginning his career.

At the tundra's edge he paused, gave thanks to God for His bounties to one so undeserving, and then, with his collecting cases slung over his shoulder and his knapsack on his back, Carl Linnaeus started off across the flower-patterned, stony-barren, lichen-covered expanse of the tundra and trudged toward a range of low, somber, barren hills. There he saw a dim line of animals that moved slowly, apparently fed as they went, and disappeared over the crest of a hill. As gray as the lichens themselves, they were exciting to see, and he wondered what they were and where they were all going with so much purpose. He did not catch up with them that day. He became too much engrossed in the wild flowers which grew in the seemingly barren landscape—in the beautiful little cushion pinks and miniature bellflowers and the tiny lilies and primulas and saxifrages which made a great and splendid garden all about his feet and as far as he could see. The tundra was not really gray and lifeless at all; it was an astonishing flower garden whose colors, at a distance, all blended with the gray lichens, the gray mist, and the remnants of old snow.

He listened to the larks singing, watched them mount up and up, singing, into the springtime sky, then drop down again to the gray moss and to the mats of flowers. Here among the low plants the larks had their nests, and so also did the pipits and the gray-crowned finches. He came upon a mother ptarmigan which was colored so much like the lichens and the rocks that he had almost stepped upon her before she got off her eggs. With an anguished clucking, she went flopping off on her side as if mortally wounded. He smiled at the poor foolish hen and bent for a moment to look at the mottled eggs which were nestled in a hollow in the moss. Then he went on again so that the bird could come back before the eggs were chilled. In spite of the fact that the brilliant sunlight had emerged from the mists, the June wind was cold and the tundra bleak.

When he caught up with the animals next day, he discovered that they were reindeer, thousands of them. Dwellers of the tundra of Lapland, eaters of the curly gray lichens, the reindeer were creatures which, like the ptarmigan, the larks, the finches, the pipits, and the tundra flowers themselves, were fitted to their stern environment.

For that matter, the whole expedition was full of glorious new experiences. Even the clouds of mosquitoes and black flies, horrible as they were to endure, were astonishing to him in spite of the punishment they inflicted. With each mile of new landscape, the world opened more and more to him and to his inquiring mind. Day after day, young Linnaeus, in his stout fur jacket and upturned Lapland boots, his flat green cap set at a somewhat rakish slant over one merry brown eye, his collecting cases soon crammed, gathered specimens of all the Lapland plants he could find. He forded rushing streams and saw wild waterfalls in the mountains, came to villages of people who had strange customs and stranger costumes. He found with something of a jolt—he was, after all, a minister's son—that many of the Laplanders were not Christians, but pagans, as the old Vikings had been before the Norse king had had them all converted whether they wanted to be or not.

After traveling 4,600 miles in five months, largely on foot, he had found 537 species of plants, more than a hundred of them totally new. One which especially pleased him was a charming little creeping plant which he had come upon in the mosses of the

Twinflower

spruce forests, along the open tundra, and in the mountains. It was such an exquisite little thing that he quite fell in love with it. It had thin, trailing, yet sturdy vines with tiny round leaves that were scalloped on the rims, and, rising from small leaf rosettes placed at intervals along the creeping stems, thready little stalks stood two or three inches tall. Each bore a pair of pink, fragrant bells. He called it twinflower.

Seven years later, Johann Friedrich Gronovius, a great botanist, wishing to honor Linnaeus's rising fame in the world of science, asked him which flower he would choose to bear his name. There was no hesitation. He would select the modest little twin-flower, the one he had found in Lapland during that happy summer of 1730 when the world opened before him. Gronovius thereupon gave to this plant the Latin binomial *Linnaea borealis,* which it keeps today. It is identical with the twinflower which creeps across mosses and over rocks and under conifers in the forests of northern North America, as common in Wisconsin and Maine as it is in Lapland and Norway. Our twinflower and Linnaeus's—it is immortalized in his portrait, which shows him holding a spray of the tiny Linnaea.

Brown as a Laplander from the splendid sunshine, Carl Linnaeus came back from the north. With the unquenchable gleam in his eyes sparkling all the brighter and with incomparable experiences to enrich both his memory and his future work, he was quite ready, immediately on his return, to go out on whatever journey the Royal Society might next choose for him.

The Royal Society indeed had been impressed with the young man. In spite of his obscure origins and his background of neither wealth nor prestige, he was evidently headed for a career in botany, and the members of the Royal Society respected him. He had still to go a long way before his name was known around the world, but he was already well started in that direction.

Today the name of Carl Linnaeus probably appears more times in more scientific books, especially in the manuals of botany, than that of any other person. He was the mentor and the guide, the pointer of the way toward a clarifying of the thick confusion of names and classifications in the field of natural history. He is usually pictured as a plump, pleasant-faced gentleman, a jolly person in a curly white wig and looking something like Haydn or Handel, with whom he was a contemporary. Yet he was once a young, hot-blooded, handsome, eager young man with a joyous sense of humor and pleasant disposition who pioneered in the field of botany and, in his pioneering, never lost sight of the first glow he knew in his deep love for plants.

Botany at that time was not a field in which a man might expect to make a good living. Even when financed by the Royal Society, a man could seldom support a family, and there were only too few positions then open for teachers of botany. Linnaeus's mother had wanted him to be a pastor, like his good father; his father himself would have been satisfied if his son had become a doctor. Carl himself might have been quite as well pleased if he made no money at all and simply had the chance to collect plants all over the world and find ways of giving them names, but then, as now, a man had to make money to live. Carl was unmarried at the time of the Lapland trip, but he knew he had to look ahead to his future. Since most science at that time was oriented to the education of a physician, he therefore studied to be a doctor.

His father, the Pastor Linnaeus, could scarcely complain if the son loved plants. The pastor himself had started the child when only a year old on the happy route of botany. At first, it was a simple matter of showing him the beauties of a garden flower; then, as the child grew old enough to talk and to understand, the father insisted that he learn the names and properties of every plant in the parsonage garden. Since Pastor Linnaeus was a dedicated collector of unusual plants, there were a good many species to learn, so that very early in his life the lore of plants and the fascinations of botany had Carl Linnaeus firm and fast in their delightful grasp.

Even his name was taken from the family's favorite tree. A large and much-beloved basswood, or linden, was held in such reverence, and was so old, that no one even dared to remove fallen twigs from beneath it. It was marvelously fragrant when it blossomed in June, and the family's beehives were always filled with linden honey. It was the custom in Sweden at that time to change one's name at will, often choosing a certain tree or special object from which to form a name. The family for many generations had been known as Ingemarsson. Now they adopted the name of the tree, some of them going by the name of Tiliander, the pastor himself changing his name to Linné, or Linnaeus, since Latinizing a Swedish name was also a common practice. So fitting an identification thus was very suitable to a man whose career was to be engrossed in plants.

Carl Linnaeus's great mission was not only to find names for all plants, but also to standardize the nomenclature and make each species uniquely and unmistakably distinctive. Each must have a name which, by means of some orderly method of plant identification, could be distinguished by anyone, anywhere in the world. Thus far, there was no such procedure. For centuries, men had been giving haphazard, nonconforming, and often casual names to plants. In the distant past, vegetation had been regarded from two very different standpoints—the one philosophical, the other purely utilitarian. The latter was an offshoot of the study of medicine or of the development of agriculture, in which a plant was thought to be of no good unless it had a use for mankind; the former was the beginning of the science of natural philosophy, which became natural history.

Botany had its beginning in Greece when Aristotle, a pupil of Plato (384 to 322 B.C.), became interested in the natural world. Considering that the "soul" of the plant did not feel sorrow or pain, he thus regarded plants as obviously lower and less important in the order of life. Aristotle's pupil Theophrastus inherited the great teacher's library and his theories, and then on his own he set about to progress a long way from the basic teachings of Aristotle.

Theophrastus wrote a book called *Enquiry into Plants,* which was concerned with the vegetation of the Mediterranean around Greece, the area he knew most about, but it also included specimens of plants which Alexander the Great had brought from distant places. Apparently Alexander was so much aware of his natural surroundings that he took with him on his punitive expeditions several men who were versed in plants and other objects of scientific interest, so that collections made in the midst of war might be used for study in Greece during the times of peace. Thus Theophrastus had access to plants from Egypt, from the northern countries, and from the Far East.

But botany as a science, after this promising start, limped along as mankind lost contact with much of the mental progress of Greece and Rome and lapsed into the Dark Ages of the mind. Yet here and there men were still poking and searching among plants, still quoting Aristotle and Theophrastus, but advancing little farther than these two botanical pioneers. Now and again, however, a man of inquiry and knowledge stood forth in the gloom.

Dioscorides in 512 A.D. produced a magnificent *De Materia Medica* with splendid illustrations, but there was little else of this sort until the fifteenth century, when the herbalists in Europe took over the science of botany and produced books and manuals called herbals. These were partly the result of the invention of the printing press with movable type, partly an outcome of the surge of knowledge welling up during the revival of education during the Renaissance. The herbals of the fifteenth and sixteenth centuries in Europe were the foundations upon which later botanists built their growing science, for in the often inaccurate herbals there nevertheless dwelt the hard core of basic truth.

In all the search for knowledge and in all the business of botany, there had never been an honest nor a totally workable method by which unknown species might be identified according to family and kind. Names given were lengthy Latin descriptions; only rarely was a plant designated by two simple Latin names. An example of the long descriptive name was seen in Johann Bauhin's carnation: *Dianthus floribus, solitarii, squamis calycinis subovatus brevissimis, corollis crenatis.* It was no doubt an excellent description of the carnation, but it was hardly an easy one to memorize or use.

When Linnaeus was working out his simplification of plant names—believing that all any plant needed were two of them, or at the most three, to be sufficiently well described—he boldly shortened Bauhin's carnation to *Dianthus Caryophyllus,* for which, and for the rest of his simplified plant names, the world shall always be grateful.

He felt that plants ought to be arranged in families whose structural similarities united them. By examining the plant's sexual structure and determining the character and number of stamens and pistils, he learned how to tell at once to which family it belonged. Petals and sepals, he knew, did not always tell the truth about a plant family. Yet, by looking wisely at the inner workings, the superficial differences of calyx and corolla merged in the unity of the similar character found in the reproductive parts of plants of one family. It was a magical key to an understanding of the plant world.

As Linnaeus examined flower after flower—campanulas and geraniums and lilies and roses, peas and dandelions and carnations and phlox, he began to see his ideas proved and clarified. The sexual system, he now was certain, was the one to use for a standardization of classification, for by their reproductive parts all plants might thus be known and arranged in the proper genera.

It was a revolutionary theory to men of science who, for many centuries, had been groping for plant knowledge. They evidently had not considered that plants even had sex; animals, yes, but not the inanimate plants. The setting of the seed, or the failure to set seed, seemed the result of a more or less divine or supernatural accident. When insects were observed to be visiting flowers, their presence was explained as their own vague business of gathering nectar, not a mission which possibly might be vitally connected with the flower and its reproduction. By opening new channels of conjecture and thought in this direction, Linnaeus had produced an explosive theory. Using as stepping stones the accumulated knowledge of men who had preceded him, he had reached the monumental conclusion that plants *did have sex,* and that pollen from the stamens, long looked upon as some sort of ornamental but useless dust, actually had a vital purpose as the male element needed to fertilize ovules in the ovary of the pistil. The varying shapes of pistils and their cavities, of stamens and their pollen, now formed the basis of Linnaeus's wonderful new system of plant classification.

When young Linnaeus had worked out his sexual system of plants, by which they could be arranged into families, or genera, he hesitantly presented his paper on the matter to Professor Celsius at the university at Uppsala, where Linnaeus was a medical student. When Celsius examined Linnaeus's offering, he was at first dubious, then enthusiastic, finally very much excited. He had copies made of the report and passed them around. In 1737, the *Genera Plantarum* was accepted and published. Suddenly, Carl Linnaeus found himself growing famous in scientific circles. At first his theories were challenged by established botanists in England, France, and Sweden, whose duty it was to doubt any new idea. That was to be expected of any new thought or proposition. He would have been disappointed if there had not been enough interest in it to provoke argument, challenge, or disapproval.

Dr. Dillenius wrote from London: "A new botanist is arose in the north; a founder of a new method, a 'staminibus et pistillis,' whose name is Linnaeus. . . . He . . . hath a thorough insight and knowledge of botany; but I am afraid his method won't hold."

Somewhat to the astonishment of Dillenius and even of the author himself, the system did hold. Linnaeus realized that it was far from perfect; that it might not remain as he first arranged it. He knew he was a pioneer, and his successors must clarify and improve his work, perhaps even find a newer and better system to supplant his. He was even now doing a considerable amount of stretching and alteration of his system to fit all the infinitely curious plants which were being brought to Europe by the collectors in far places of the world. These foreign species were the supreme tests of the method and, if it needed revision at times, that made it all the stronger. He always believed that his sexual system was only a stepping stone to a better method which might more easily fit every flowering plant in the world.

Imperfect though his method might thus have been in spots, Linnaeus had nevertheless revolutionized botany. With his *Genera Plantarum* he opened the way. In 1753, his *Species Plantarum* marked the official start of binomial nomenclature upon which all of today's plant names are founded.

Carl Linnaeus never reached the distant lands to which, in his exuberant youth, he had longed to go. When next he was sent on a collecting expedition by the Academy of Sciences, in 1734, it was only to Dalecarlia, a province of Sweden, immensely dull as compared with the jungles of Brazil, but immensely safer than such an expedition might have been. It was no doubt a better thing for a botanist of growing note to remain in Europe, make his field trips no farther than Russia, and work at classifying the bales of plant specimens which more reckless yet perhaps more fortunate explorers were sending back to Sweden, France, Germany, and England.

He was very busy. With his *Genera Plantarum* accepted by

botanists, he was given the post of professor of botany at the university at Uppsala when old Dr. Olof Rudbeck died. Linnaeus married a pretty girl he had met in Dalecarlia. He had a full life and, as his favorite recreation, often went on botanical excursions with his students, always coming home refreshed from these trips and with plants which he himself had collected. He never lost his enthusiasm for them, or the challenge of finding suitable names for every new one which appeared. He must have had great fun in doing it.

For some plants he could simply retain the old classic names which had been in use since Aristotle's and Theophrastus's time; such comfortable and familiar epithets as *Rosa, Lilium, Quercus, Betula,* or *Fagus* for the rose, lily, oak, birch, and beech were still eminently suitable. There was no need to change these, nor any wish to do so. They were a sound basis on which to build other names. He used some of the classic Greek and Roman names of the gods and goddesses, nymphs, queens, or mythological figures for some of the more exquisite heaths and orchids—the *Cassiope, Andromeda, Pieris, Calypso, Zenobia, Arethusa,* and *Daphnae.* He named some plants for the men who had discovered them, or who were associated with him, or who were noted elsewhere in the field of botany, such as *Rudbeckia* for good old Dr. Olof Rudbeck, who had done so much for him, or *Bartonia, Claytonia,* or *Mitchella* for botanists in America who were sending specimens to him for identification. Or, he might coin hybrid words by combining Greek and Latin roots, or else create some seemingly complicated and delightful ones such as *Arctostaphylos Uva-ursi,* for the bearberry, to describe its clusters of pearly, pink-tipped bells and red fruits which he fancied as tiny bunches of grapes that were relished by bears.

Specific names were easier to come upon than family names. A plant might have a specific name which commemorated its discoverer, such as *Lilium Catesbyaei,* or it might indicate the place, the time of year, the texture of the leaf, the color, or some other characteristic of the plant. Within the scope of one word he endeavored to condense the meaning of two dozen words as presented in the lengthy Latin descriptions of the past. It was a mighty challenge, yet many of them really caused him but little trouble. Certain New World plants could so easily be designated as *canadensis, americana,* or *virginiana.* A northern species was *septentrionalis,* a southern one *australis;* while places, such as New York, Florida, or California, where a plant might have been discovered, could be called *novaboracensis, floridana,* or *californica.* Colors were easy—*rosea, rubra, coerulea, violacea, viridis, alba.* He could indicate characteristics of growth by using such terms as *prostrata, erecta, minima,* or *gigantea,* or the time of bloom such as *verna,* for spring, or *hyemalis,* for winter. It was neat. It was compact. It was workable. *Above all, it was workable.*

Linnaeus's Latin names were excellent, but it was not always possible or desirable for everyday people, on everyday occasions, to use the botanic terms. The better known a plant was, the more common names it had acquired. These were apt to vary, besides, in certain parts of the country. It was, in fact, sometimes possible to trace a person's background, especially in America, by hearing what he called certain plants. Puckerbrush was a New England name for the bayberry; candleberry was another; North Carolinians, however, called it wax myrtle. In the South, the wait-a-bit thorn was the name for the common green brier, but in the Midwest it was catbrier, and in some places it might be called hellfetter. The Amelanchier, or shadbush, became known variously around the country as service berry, sarvis, Juneberry, sugarplum, and saskatoon, depending upon whether one lived in the Northeast, the Middle West, or on the Northwest Coast and in Alaska.

Confusion lay in common names. To one person the adder's tongue might be a kind of fern, to another a plant also called dogtooth violet. The dogtooth violet, however, is not a violet, but a lily. By calling the fern an Ophioglossum and the lily an Erythronium, all question is removed and the usefulness, clarity, and the downright necessity for the Latin names is made even more clear than before.

The cool, calm Latin names as designated or created by Carl Linnaeus and his successors remain steadfast. They are a clearing not only in the jungle of overlong polynomials of the past, but in the multiplicity of common names of the present. When Linnaeus stabilized nomenclature, he made it possible for naturalists and botanists around the world, no matter what language they might speak, to understand each other. This was the great mission of Pastor Linnaeus's little boy, who began as soon as he could walk and talk to learn about plants in his father's garden.

Franklinia

3. John Bartram, the King's Botanist

There was no reason why John Bartram should have become more than what he was, a conscientious Pennsylvania Quaker farmer who believed in the goodness of the land and in the work to improve it. Born in 1699, he had had only the usual education of a workingman's son and had traveled little. For a long time his greatest ambition was to rear his family, raise better crops, and discover ways in which he might increase the usefulness and fertility of his farm at Kingsessing, along the Schuylkill, not far from Philadelphia. He loved the land and he loved plants, but aside from the seeds he planted and cultivated, he apparently knew very little about what grew around him or about the subject of botany itself.

But something happened to John Bartram. It may have been a gradual thing, an emergence from groping to positive action, for he was certainly not the ignorant, totally unaware man which a pretty story, now a legend, relates of him. The tale goes that on a hot day, about the year 1730, he had left his plow and had retreated to the shadow of a large tree to rest for a while. As he seated himself and stretched out his legs, he saw a daisy. He had been seeing wild daisies all his life, but until now, so the story goes, he had never really noticed them. Some perception which he had not used before let him see this flower as with a mental magnifying glass. He picked and examined the blossom; and although he knew little of names or the significance of the flower parts, he did know that in his hand was a marvelous little entity. Its distinct parts were all arranged with an obvious order and beauty, with a meaning which he was impelled to try to comprehend. There were white rays, which he may have called petals, arranged like a small sunburst around the golden center. And that yellow center, he could now see with his mental spectacles so well adjusted, was not at all the flat yellow cushion he had always supposed it to be. It had parts of its own. It was, in fact, a marvelously compact arrangement of slim golden tubes out of which emerged small yellow threads and dusty little knobs. He turned the flower over, saw how the whole creation was set firmly in a green cup, and that the cup itself was wonderfully well ordered also. That daisy—it was the key to the great doorway opening to a whole new world.

Later—much later—when he was a famous botanist and someone asked him the usual question put to so many naturalists—how he first came to be interested in plants—he is said to have replied, in his quiet Quaker manner:

" 'What a shame,' said my mind, or something that inspired my mind, 'that thee shouldst have employed so many years in tilling the earth, and destroying so many flowers and plants, without being acquainted with their structures and their uses.' "

Perhaps the story is only a story, a legend created around a man who was one of the early successes of American history and science. It matters very little now. Many individuals, at some time in their lives, as if looking at a daisy through a hand lens and really seeing it for the first time, may reach a sudden moment of revelation. Many a person has come suddenly to an awareness of the values of nature and to the satisfactions to be found in searching for and understanding flowers, birds, insects, or trees. Yet there had had to be some moment of awareness in John Bartram in order that his life's course might be influenced so strongly that he would often leave his farm to his servants and his family and, for the remainder of his long life, go off periodically on plant-collecting expeditions. It was not the customary procedure of the dedicated American farmer in the eighteenth century.

In America there were as yet few adequate means by which a man might learn about his surroundings. Since the great advances in botany were taking place in Europe, John Bartram had to do a considerable amount of groping and laborious searching among the extant plant lore in order to find names for his trees and flowers, or else had to make his own names. It was a day of great excitement for the whole family when John Bartram triumphantly brought home from Philadelphia a copy of the revolutionary new *Genera Plantarum* of Linnaeus. He spent long evenings in ferreting out its meanings. It was of course all in Latin, and since his labored schoolboy Latin was in need of a considerable amount of refreshing, he may have had to go to the neighborhood school teacher for help. However he managed it, John Bartram mastered the Linnaean system and then set about identifying every plant on his farm. He was charmed to see how Linnaeus's method applied to all he found. In a way hitherto impossible, it put plants neatly into their families.

Then, to find ever more specimens, Bartram began a series of botanical excursions into the countryside to seek out new ones and, whenever possible, to bring back plants, seeds, or young trees for his garden. At the same time, he taught botany to his children, one of whom, William, also became a noted botanist. Until the boy was old enough to travel in the wilderness, John Bartram left him at home with the others in his wife's capable and understanding care.

John Bartram traveled to Maryland, into Delaware, and to the mountains of the Blue Ridge; he took a five weeks' trip to the southern Appalachians; he traveled north with Conrad Weiser, the Indian fighter, to Lake Ontario and the Mohawk country. Bartram's collections of living plants expanded. Before he had quite realized it, he had created the first botanical garden in America.

When he ran into insurmountable difficulties in identification, he made bold to write to Dr. Linnaeus himself (all in Latin, as was the custom of polite science of that day, and very much labored over by the farmer of Kingsessing before he was satisfied with it) and painstakingly explained some of his problems. To Linnaeus he also sent packets of plants neatly pressed between papers, and the great man wrote back in a growing enthusiasm for the excellent specimens which the man in Pennsylvania had the perception to find and send to him.

Quietly, John Bartram grew famous. Other botanists, hearing about him and what he was doing on his farm, came to see him, to examine both his farming methods and his collections, and to walk in the splendid botanical garden. Some of them brought plants, cuttings, or seeds as gifts, which he planted in gratitude. He always sent along some samples of his garden with his visitors when they went home.

It was also customary for botanists and naturalists to carry on a long correspondence with each other. It was one of the few media for exchange of ideas and discoveries at a time when scientists were all too few and scientific literature even more scanty. By writing to each other—often never meeting—and exchanging bundles of pressed specimens, living plants, or seeds, both in America and abroad, they gave to American botany, as well as to botany of the world, a tremendous impetus.

Among John Bartram's friends by correspondence was Peter Collinson, a Quaker merchant in London who also had a hand in assisting several botanists who came to America to collect. Collinson had a garden of his own to which he yearned continually to add new specimens from America. He and Bartram had a long and happy exchange both of letters and of plants and seeds. Each man coveted species growing in the other's garden or, in the case of the London merchant, in the wild parts of America. Collinson was becoming more and more impressed by what the man in Pennsylvania was accomplishing in discovery. He felt that something should be done to assist Bartram, help to finance the field trips from which he was bringing back so many rare finds, and permit him to work at botany without having to worry about supporting his family. The farmer of Kingsessing might then engage more men to help him in the fields, orchard, and botanical garden, and could roam freely in quest of more plants.

Collinson said nothing of what he was about. While the mills of change turned at their own deliberate pace, John Bartram continued to collect and study. He now wished to visit parts of the country which he had never seen. In 1763 he arranged to be away long enough to go with his son William, who was then twenty-five, into New York State to visit one of his correspondents, Dr. Cadwallader Colden. The wealthy Coldens lived on a large estate near the Hudson River, but since Cadwallader and his daughter were botanists, it was this bond which made farmer Bartram and New York's Surveyor-General, Dr. Colden, of one mind and interest. It was glorious to go out collecting at dawn and ramble with kindred spirits all day in the Catskills, and, during the long and delightful evenings, avidly discuss botany while arranging the specimens on papers. There were all too few men in America who could discuss botany sensibly and with intelligence.

Miss Jane Colden could also discourse with authority. So, of course, did William. The gathering was made even better and more stimulating when, one day during the Bartram visit, Dr. Alexander Garden of Charleston actually came on the long journey all the way to Coldengham near the Hudson to join in the field trips and the inspiriting botanical conversations. Dr. Garden, for whom the gardenia would be named, urged the Coldens and the Bartrams to visit him in the South. Colden thought it unlikely that he himself could get away, but the Bartrams lived much closer to the Carolinas, were fairly footloose, and old John even then had a great yearning to visit the fascinating South and see Florida. Perhaps next year he would be able to afford it.

But when, in 1765, John and William Bartram did make that long-anticipated journey to Florida, it was under extraordinary circumstances. That was a remarkable year, not the least of its wonders being the tremendous honor which came at last to John Bartram. Collinson and others had worked very hard in his behalf, and now he received a formal notification from the King of England himself that he had been appointed Royal Botanist in the colonies. George III was often called Farmer George because of his love for plants and gardens; he had a rising botanical garden of his own at Kew, for which he now coveted new specimens from America. Bartram, in thus aiding the king and being paid for it, could now travel more extensively. The stipend allotted him would finance all those trips which he had been taking on his own, or which he would hope to take before he grew too old.

The event put quite a glow on the household. Their friend Benjamin Franklin congratulated him. Letters came from friends in America and in Europe. Royal Botanist—plain Quaker John Bartram had the feeling of disbelief which comes to a man who has worked all his life and has obtained his life's dreams only in the long, hard way of personal labor. If the honor had come to him when he was a young man, perhaps it might have gone to his head. As it was, he was now sixty-six, and although he appreciated both the honor and the money, they were not earth-shaking.

Much better was that expedition on which he and William set out at last in the summer of 1765, bound for Florida. The king had commissioned him to explore the St. Johns River, discover its source, and send some handsome new Florida flowers and trees for the royal garden and greenhouses.

The Bartrams and their collecting equipment went by ship from Philadelphia down the coast to Charleston, where they were met by the cordial Dr. Garden and escorted to his comfortable home beneath spreading live oaks hung with Spanish moss. The Bartrams were delighted with the moss, with the oaks, with the semitropical appearance of the palmettoes, the figs, the hibiscus, the oleanders and pomegranates, the lush manner in which the roses and honeysuckles grew—all were a delight. It was a botanical adventure simply to wander around the streets of Charleston and to look over iron fences into gardens. But their host had other places and other plants to show them. They ranged out into the back country. They tramped the live oak wilderness and waded into swamps where spider lilies grew. They found big pink Sabatias and lavender Thalia, and incredible thickets of tall green cane where wild razorback hogs lurked and unknown birds sang.

After a summer which had been both hot and rainy, John and William Bartram at last left Charleston late in September and

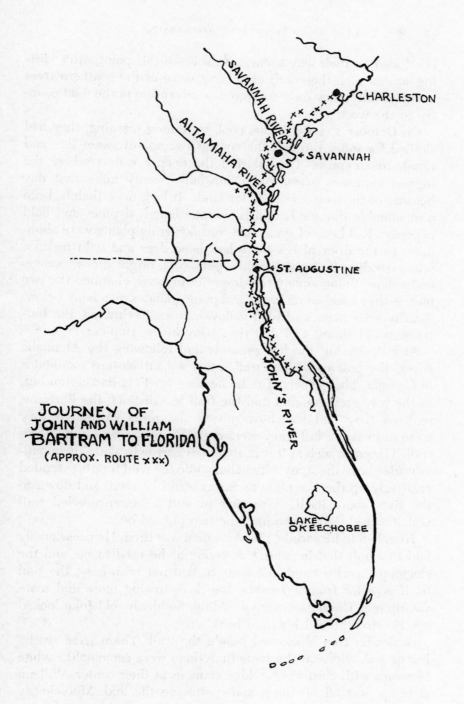

CHARLESTON

SAVANNAH RIVER

ALTAMAHA RIVER

SAVANNAH

ST. AUGUSTINE

ST.

JOHN'S RIVER

JOURNEY OF
JOHN AND WILLIAM
BARTRAM TO FLORIDA
(APPROX. ROUTE xxx)

LAKE
OKEECHOBEE

continued on their way to Savannah. From this point, after visiting several men there and examining some of the southern trees
in the vicinity, they rode out into new adventure in the wild country to the west.

On October 1, 1765, a fine, cool, refreshing morning, they had
sloshed for miles through the coastal swamps of sweet bay and
tupelo and cypress. They skirted the deeper waters where the
biggest cypresses grew, and made only twenty miles that day
because of the wetness of the lowlands. It had, nevertheless, been
a memorable day for botanists, for the long-leaf pines and bald
cypresses had been of great size, and flowering plants were abundant. In the drier places there had been deer and wild turkeys,
while dazzling flocks of Carolina parakeets, bright green, orange,
and yellow, flying among the draperies of moss, charmed the two
men as they lunched on bread and pomegranates on a bank below.
Nearby were masses of wild yellow cannas. Enthralled, the Bartrams could almost feel that they were in the tropics.

At last, leaving the low grounds and following the Altamaha
River, the pair rode up a trail into the southeastern mountains
of Georgia. They expected to be heading for Fort Barrington but,
as the way grew rougher and the trail less distinct, the Bartrams
realized they had somehow missed the route. Although they
were not exactly lost, they were certainly following an unknown
trail. The path, such as it was, mounted narrowly along the mountainside above the stony Altamaha, while the wild forests extended
massively up the mountain as far as could be seen, and down to
the river shore itself. Expecting to find a better-traveled trail
which would lead to the fort, the two pushed on.

Hunched in his saddle, the older man was tired. He occasionally
had to admit that he wasn't as young as he used to be, and the
slogging travel through the swamps had not been easy, the trail
itself now far from agreeable, the day growing more and more
wearisome. The horses were plodding. Suddenly, old John looked
up. He straightened his tired back.

A slender tree blossomed beside the trail. There were scarlet
leaves and, oddly, at the same time there were camellialike white
blossoms with clusters of golden stamens at their center. William
rode up, was off his horse and examining the find. Marvelously

revived, John Bartram dismounted with care and evaluated the specimen. Both thought at first that it was simply a Stewartia, that wild camellia of the South, in bloom out of season. But when William and John with their skillful botanical eyes pored over this discovery, they knew it was not a Stewartia. In all their reading and their discussion with other botanists, they had never known a tall, slender tree like this, one with smooth gray bark which covered the sinewy trunk and boughs like the sleek skin of a greyhound. They could not recall a tree which blossomed so profusely and splendidly at the time when the tree itself was blazing red in autumn. It was a puzzle. It was one of those delightful conundrums which naturalists love to find and which set lanterns of glory along the trails to adventure.

Adding leaf and flower specimens to the saddlebags, the pair wished there also were seeds, but they could find none. It was scarcely possible to take plants, they thought, but, since it was so late in the season and much vegetation was approaching dormancy, they chanced it and took along some seedling plants anyway, on the possibility that at least one might live.

These unknown seedlings from the wilderness of the Altamaha eventually arrived in Pennsylvania and grew in the Bartram garden. One blossomed when it was only three or four feet tall. Since the tree was indeed a new and unknown genus, John Bartram named it for one of his best and closest friends, Benjamin Franklin, calling it *Franklinia Alatamaha*. People who were interested in trees came to the Bartram botanical garden to see the rare new Franklinia, but no one, not even the Bartrams themselves, made the long trip back into southeastern Georgia's rugged mountains to the place where the rare trees grew. Not until 1790 when William Bartram was on his way home from visiting the Creek Nation did he make a side trip to obtain more specimens and, he hoped, seeds of the Franklinia. He managed to find the place where he and his father had come upon it by chance in 1765, and secured what he wanted.

The Franklinia readily adapted to cultivation and was easily grown by people fortunate enough to have been given slips, seeds, or seedlings. But, after that second journey to the tree's original habitat in 1790, no one ever again saw the Franklinia growing in the wild. The tree in its secret haunts since then has dropped from the knowledge of men. It has continued to do well in cultivation. Nurseries long ago took over its propagation and it is available today at a moderate price from several firms in the East. But never, since 1790, has that strange, lovely, lost Franklinia been seen outside cultivation. Each generation of botanists has had men in the field. They have combed the slopes above the Altamaha and the mountains of southeastern Georgia; they have puzzled over how so attractive and prominent a tree could possibly have escaped the keen eye of the plant hunters. Nevertheless, all of the Franklinias in existence today—unless the tree still hides somewhere in the mountains of Georgia, or was rediscovered only yesterday afternoon—are descended from those original plants brought to Kingsessing by the Bartrams in the eighteenth century.

Following their discovery of the new tree—which, at the time, did not seem to be as momentous a find as matters later developed—the two rode on and at last found the trail to Fort Barrington. After several days there, they continued their journey and eventually reached St. Augustine, Florida. On their way they

Yellow Pitcher Plants

found more new and fascinating plants. In low sandy fields and marshes, the tall, upstanding trumpets of the yellow pitcher plant were exciting. They were not new to science, but were new to the Bartrams. William wrote ecstatically:

How greatly the flowers of the yellow Sarracenia represent a silken canopy. The yellow, pendant petals are the curtains, and the hollow leaves are not unlike the cornucopia or Amalthea's horn. What a quantity of water a leaf is capable of containing: about a pint! Taste of it—how cool and animating—limpid as the morning dew.

The carnivorous plants fascinated the Bartrams. When they came upon bogs where the Venus's fly-trap grew, they thought it was surely one of the most wonderful plants in the world.

But admirable are the properties of the extraordinary Dionea muscipula! . . . Astonishing production! See the incarnate lobes expanding, how gay and sportive they appear! Ready on the spring to entrap in-

Venus's Fly-trap

cautious, deluded insects! What artifice! There, behold, one of the leaves has just closed upon a struggling fly; another has gotten a worm; its hold is sure, its prey can never escape—carnivorous vegetable! Can we, after viewing this object, hesitate a moment to confess that vegetable beings are endued with some sensible faculties or attributes, familiar to those that dignify the animal nature?

In Florida, by December, they were much interested in the pale green rafts of the water lettuce (Pistia) floating on the quiet reaches of the St. Johns. The water lettuce was part of a well-populated habitat of its own, which included the haunts of the alligator and the otter, a place where frogs sat, and where crows and gallinules came down to walk about and forage, as if the unstable footing of the plants indeed made solid islands. The water lettuce beds seemed so complete as a natural habitat in itself that William Bartram said whimsically: "There seems, in short, nothing wanted but the appearance of a wigwam and a canoe to complete the scene."

With gratification they saw that almost everything they found in Florida was new to them, or, even better, new to science. They discovered the oil nut, or *Pyrularia;* the royal palm, *Roystoneia;* the yellowroot, *Zanthorrhiza.* Along the St. Johns River they found magnificent magnolias whose glossy evergreen leaves were backed as with brown velvet, the upcurving branches of the large trees bearing enormous, fragrant white blossoms which, to the Bartrams, embodied the great beauty of the South. Indeed, part of their mission had been to look for magnolias; since none was native to Europe, these lovely trees were much admired as garden prizes and much sought after as more and more were being discovered in America. Bartram sent a number of them to his friend Collinson and to the king. Of them all, the most admired by all, including the Bartrams themselves, was that unutterably fragrant, evergreen-leaved species, *Magnolia grandiflora.*

On the St. Johns, the two went by canoe and dutifully and with growing excitement followed the river's winding course from the sea to the source, as the king had wished. They saw with wonder the palmetto country, the grass marshes, and the prairie, all drenched in winter sunshine and alive with herons, cranes, ducks, geese, and other waterfowl. They saw the long-leaf pines and the bald cypresses and the Spanish moss, and some of the most charming and beautiful wild flowers in America. They brought back specimens of everything they could find or transport, including seeds, roots, and bulbs. A shipment of plants went directly to Collinson in the hope he would receive them all safely.

By the spring of 1766, old John Bartram was back home again. It was time for a farmer to get his crops in and see to the multitudes of jobs which were awaiting him on the farm and in the botanical garden. William, however, very much against his father's wishes, had not returned. Bill Bartram had fallen in love with Florida and he stayed there. He planned to become proprietor of a small store beside the St. Johns, later intending to buy a plantation. In reality it must have been the wildlife rather than the business opportunity which kept him in the South, for dreamy, impractical William Bartram's store soon failed, his plantation did not materialize, and he again took to the wilderness before finally coming home for a while to his father's house.

Before that, John Bartram had written to his friend Collinson about the Florida experiences:

I have brought home with me a fine collection of strange Florida plants which perhaps I may send sometime this summer, some for the king and some for thyself, but I want to know how those I sent from Charleston and Georgia is accepted, or those I sent last spring for the king from home. . . . I hope what specimens I sent for thyself will give thee great pleasure, as many of them is entirely new, the collecting of which hath cost thy friend . . . pain and sickness which fret me constantly near or whole two months in Florida. . . . Yet somehow or other I lost not many hours time of traveling through those processes, and when at Augustine with fever and jaundice, I traveled both by water and land around the town for many miles. . . .

Obviously, it would require more than illness and age to stop John Bartram. Plants were his life, his delight, and his satisfaction. His botanical garden was infinitely precious to him; it represented the results of a lifetime of plant collecting. Each plant had its personal history, its joy, and its place in his life. Thus, when the American Revolution was taking place, and, more than once, Philadelphia, the national capital, was threatened, old John Bartram grew sick with fear—fear that was not for himself, but for that other self, his garden. As a Quaker, he could not endure the thought of war; as a gardener, he could not face the thought of senseless destruction of his plants, his work, and his dreams. He worried; he grew thin; anguished lines altered and aged the placid face of a man to whom nature had given only peace and assurance.

Times grew worse. The British had taken New York, were advancing from several directions toward Philadelphia. When George Washington with his ill-trained men met General Howe on Brandywine Creek and the Americans fell back in defeat, John Bartram knew with a sense of growing horror and black fatality that his beautiful, tree-shaded garden would be next. They would overrun Philadelphia, and then . . .

Less than two weeks after the defeat of the American army at the Battle of the Brandywine, John Bartram died. He was seventy-eight years old and essentially strong and hearty, but it had been the killing fear, his relatives said, which had destroyed him as surely as if he had fallen with a bullet in his chest. And in spite of the sacrifice of John Bartram, the British did not, after all, surge into and ruin the Bartram botanical garden. He was still the King's Botanist and perhaps the British army respected him for that.

William Bartram, who lived until 1823, dying at the age of eighty-eight, all his life pursued the collecting and knowledge of plants and wild creatures. By the time William Bartram had left off his study forever, American botany had come a very long way from that moment in time when John Bartram, Quaker farmer, suddenly and with inner illumination may have focused eyes and mind upon the wonders of a flower.

Goldthread

4. Jane Colden, First Woman Botanist

In the eighteenth century, few proper and well-brought-up young ladies would have been found roaming the woods, the mountains, or the cliffs above the Hudson River. Few, in fact, would have had the wish or the reason to do so, since it was not only unseemly, but also a downright dangerous pursuit. There were still numerous Indians, often hostile ones, in New York State in the mid-1700's, so that for this and other reasons it was far from safe for a girl to go about casually in the woods. Apparently oblivious to danger or to gossip, however, Jane Colden walked the woods, climbed the cliffs, and waded the swamps. No doubt the matrons of Newburgh gossiped and perhaps her mother worried, but since Jane was a strong-willed, rich, and intelligent girl, she apparently let neither gossip, Indians, nor parental concern deter her outdoor career.

Jane Colden knew what she wanted. For a full life, she had no use for an existence which was limited to staying at home, or of promenading under a parasol, or of performing in polite society.

Although she evidently became a competent housewife, she also was an expert botanist. She wanted plants, as many plants as she could find, as many as she could find the names for, and, if she could not manage to identify them all, she would send them to Dr. Linnaeus in Sweden, who would help her. In her search, she soon exhausted even the vast acreage of the family estate called Coldengham. Then, sometimes taking a servant, sometimes accompanied by her father (who also was a botanist), she rode out with her baskets and her plant press on the trail to the Shawangunk Mountains to the west, or through the Hudson highlands, or down the road toward West Point, or perhaps ten miles east to the great cliffs margining the Hudson River, where the duck hawks nested. As America's first woman botanist, Jane Colden, during her short but exciting life, pursued her career with a fierce dedication and a great joy.

She had been reared in a family which had come to New York from Scotland, a family to whom education was a tradition, not a luxury, assuredly a benefit intended for both men and women alike. Her father, Cadwallader Colden, a graduate of Edinburgh University, came as a physician to America not long after becoming a doctor. Born in 1688, Colden lived until 1776, a fitting enough time for the staunch old Loyalist to leave the world. He had been continually infuriated at the insolence of the rebels in the colonies who had not only refused to pay the stamp taxes, but had openly defied the king in starting a rebellion which turned into a revolution. Colden, in his official attempt to enforce the Stamp Acts in New York, had been nearly mobbed. He was a politician, a king's man, who believed that to make America strong it should be more closely kept as a part of England and have a parliament and other accouterments of British rule. Being so far from the Mother Country and away from direct supervision of the king had given to many of the colonists a regrettable feeling of arrogance and independence.

Colden was one of the best-informed men in the colonies. From 1720 until 1762, he had been first surveyor-general of the colony of New York, and lieutenant-governor from 1762 until his death. He was a historian and author of a history of the Five Nations of the Iroquois (published in 1727), but botany was his greatest love. He became the first resident botanist in the colonies to make a serious study of the myriads of plants which grew in the New York area. He contrived somehow to practice medicine, carry on his government duties, father ten children, and manage an estate of three thousand acres near Newburgh, at the same time becoming a botanist with an extensive correspondence with other botanists in America, England, and Sweden. With his handsome wife's help, he educated his children at home, for there were no schools close by for them. Even later on when the boys could go away to school, the girls continued to be educated at home by their competent parents. There was in the colonies at that time no adequate or challenging education for females. Mrs. Colden, a tall, pretty, dark-eyed woman, also taught the girls the housewifely arts. Jane became especially noted for her excellent cheese and butter, which she sold in the neighborhood.

Jane was four years old when the family bought the land west of the Hudson River, about ten miles west of Newburgh, New York, and set about making a home and farm in what had been lately untouched wilderness. It was a plant-wealthy region, exactly what Cadwallader Colden wanted. He must have tried to imbue in all his children his own love for plants, and no doubt succeeded, but it was Jane, when she was in her teens, who had become enthralled with the subject of botany. Fortunately for the botanical Coldens, as well as for people like the Bartrams, Linnaeus's revolutionary *Genera Plantarum* was available in America soon after it was published in 1737. Colden, who ordered new botany books from Europe as soon as he heard of them, secured a copy of Linnaeus's work as soon as it could be sent to him. In mounting excitement, the doctor forsook politics and patients while he pored over the book and compared what he read with what he saw in the flowers brought from his woods and fields. It was a fascination to master the new system, to memorize the terms, and to explore the unknown continents of flower structures. Begging to be taught what he was learning with such avidity, perhaps Jane joined him then.

Since the girl had not learned very much Latin, she could not comprehend the complex Linnaean language, so her father indulgently translated it to English, using two or three words to explain the meaning of one, and for the first time simplifying the Linnaean system of plant identification so that the girl could pore over and memorize it.

She was excited, and she was extremely capable; she soon mastered the system and was busily applying it to all the flowers she found. Collecting flowers and leaves, the father and daughter made long, delightful excursions afield, and then, in the evenings, had to be urged to blow out their candles and go to bed, so late did they sit up to work out the plant classifications before the blossoms withered.

Jane was about eighteen when her father sent to Linnaeus a description of nearly four hundred plants which he had collected in his vicinity and identified by the Linnaean method. Some which he could not identify he asked the favor of Linnaeus to look at and reach a conclusion for him.

Evidently the great botanist was impressed. A correspondence sprang up between the two men. When Linnaeus published Colden's list in 1743, this meant that those species which Colden and his daughter had been first to find would bear the names which they themselves had bestowed. In deference to the indefatigable doctor of New York, Linnaeus named a new genus *Coldenia* after him. Related to the heliotropes, it was a small plant in the Borage family.

To have a list of plants published in Europe made Colden a recognized botanist, not simply an unknown back-country picker and presser of flowers. With Linnaeus honoring him, many well-known botanists, scientists, and travelers from abroad came to pay their respects at Coldengham. The Colden family was a happy one with a large welcome for all.

One of the visitors, not a botanist but simply an old friend from Scotland, wrote afterward:

His daughter Jennie is a Florist and Botanist, she has discovered a great number of Plants never before described and has given their Properties and Virtues, many of which are found useful in Medicine, and she draws and colours them with great beauty. . . . N.B. She makes the best cheese I ever ate in America.

More and more, Cadwallader's friends, particularly the botanists, were impressed with Jane's knowledge. It was most unusual for a female to show much intelligence beyond housework and cookery, or an ability expressed in a little French, water-color painting, embroidery, and such. This remarkable girl obviously had a keen mind which had actually encompassed the complications of the Linnaean system; she could learnedly discuss botany with any man, and this may have been somewhat disconcerting to some of them. Others frankly admired her. John Bartram respected her highly and wrote many letters to her on serious botanical subjects. In one he said:

Respected friend Jane Colden:—

I received thine of October the 26, 1756, and read it several times with agreeable satisfaction; indeed, I am very careful of it, and it keeps company with the choicest correspondence,—European letters. . . .

I should be extremely glad to see thee once at my house, and to show thee my garden. . . . I showed (my son Billy) thy letter, and he was so well pleased with it, that he presently made a packet of very fine drawings for thee, far beyond Catesby's, took them to town, and told me he would send them very soon.

Alexander Garden of Charleston wrote to his friend John Ellis in 1755: ". . . not only the doctor himself is a great botanist, but his lovely daughter is greatly master of the Linnaean method and cultivates it assiduously."

Ellis was impressed too. Writing to Linnaeus to suggest that a plant be named for the remarkable Miss Colden, he pointed out a new little flower which Jane herself had discovered in the cedar forests. It had low, glossy, three-parted, evergreen leaves. Slender, thready stems two inches high bore a single white, starry blossom on each; the plant had bright yellow roots like little gold wires. Colden had sent a specimen to Dr. Fothergill in London, and John Ellis had seen it there. Knowing that another specimen had indeed been sent to Linnaeus for a name, Ellis hastened to suggest that Linnaeus thus honor Miss Colden by calling it *Coldenella*. But Ellis was too late. Linnaeus had already classified it as *Helleborus*.

Occasionally, however, even that great botanist made mistakes, and this time he was indeed in error. When the confusion was later discovered, another botanist, Richard A. Salisbury, renamed the plant *Coptis,* for its three-parted leaves. Salisbury may have had no knowledge of Jane Colden or her discovery of the plant. Thus, the little northern goldthread, boreal creature of the cold cedar woods and bogs of the upper parts of the country, found by a young lady hunting flowers in 1758, has come down to us as *Coptis groenlandica.* It is a pity, rather. Jane Colden deserved to have a plant named in her honor.

Others felt the same way; they pursued the point. Peter Collinson knew about Jane and also wrote to Linnaeus about her: "I have but lately heard from Mr. Colden, he is well; but, what is marvellous, his daughter is perhaps the first lady that has so perfectly studied your system. She deserves to be celebrated."

But she was not, and one wonders why. Perhaps it was too great a departure from the masculine world of botany at that time to give a woman's name to a plant.

No portrait evidently was made of Jane Colden, but judging from the extant likenesses of her parents, she must have been an attractive girl. Her father was red-haired and blue-eyed, her mother dark, and, since girls often favor their fathers, perhaps Jane also was red-haired and blue-eyed.

She did not sit at home waiting to have a plant dedicated to her. She was too busy scouring the great acreage of Coldengham and much of the countryside round about. During one delightful summer, she had the companionship of Samuel Bard, a boy of fourteen, who was spending several months of 1756 with the Coldens. The boy had been ill and his father, a close friend of the Coldens, begged that he be allowed to come out to the country to regain his health. Naturally Sam, curious as to what Jane was doing, trailed along after her. Then, becoming infected by her own lively interest, he volunteered to help. She had him with her on all her excursions that summer. He carried her baskets of specimens, climbed hills to obtain plants, scrambled over ledges to secure ferns. All the while, Sam Bard was learning about plants.

The woman and the eager boy, whose thinness was filling out and whose color was improving—Miss Jane's excellent cheese and butter were no doubt doing good work, too—had delightful weeks together. Samuel Bard's later career in botany was laid entirely on the doorstep of Jane Colden, to whom he remained devoted the rest of his life.

With young Sam's help, Jane collected four hundred species of plants. Of many of them she made careful leaf prints with printer's ink, and sketched many rather simple ink drawings overlaid with a color wash. She wrote accurate and detailed descriptions of all of them, one page to a flower. Her handwriting was good, but her spelling was influenced by the inconsistencies of the times. The portfolio of manuscript and drawings was completed in 1758.

The following year, when she was thirty-five, Jane Colden was married. In that period in history, a woman of thirty-five was middle-aged and was usually beyond all hope of a husband. It is to be wondered if she had had many suitors in her youth, or if she was always so interested in botany that she could not be bothered. Perhaps she was waiting for a man of like interest, and these were scarce; again, perhaps the young men of Newburgh and New York City preferred a woman who was not so frighteningly intellectual. At any rate, in 1759 she married Dr. William Farquhar, a widower who was older than she, and with whom she was evidently deeply in love. Seven years later when she was forty-two, she bore a child which died not long after birth; Jane herself died a few weeks afterward.

Jane Colden who had botanized in the woods of New York was gone, but the career of her portfolio of four hundred drawings, leaf prints, and descriptions of American wild flowers and trees was far from ended. The collection began a long and curious peregrination which is still very much wrapped in mystery. Much of the story can only be conjectured. For instance, will it ever be known how Captain Frederick von Wangenheim, a Prussian serving in the Hessian regiments hired by the British during the American Revolution, laid hands on this treasure? Cadwallader Colden was dead, and most of the children were grown and married. Somehow, Captain von Wangenheim secured Jane Colden's botanic manuscript. Since he himself was a forester and botanist and apparently was deeply interested in American trees—he is credited with having found and named our pecan, bitternut, pignut, post oak, scrub oak, and many more—he undoubtedly appreciated this piece of booty.

He must have taken it back to Germany with him, for it subsequently fell into the hands of Professor Godfrey Baldinger at the University of Göttingen, then was taken to the University of Marburg when Professor Baldinger taught there. Then—but how or why is not known—the manuscript reached Sir Joseph Banks in England. Banks was an ardent botanical explorer, had supported other explorers on world-wide expeditions, and was president of the Royal Society of London for more than forty years. When Sir Joseph died, Jane Colden's botanic manuscript found its way at last into the collections of the British Museum in London. It is there today.

Feeling, however, that this was a loss to America—a truly American product like this should not repose so far away from home—the Garden Club of Orange and Dutchess counties in New York recently arranged to publish a portion of Jane Colden's illustrated manuscript. In book form, it was first made available to the American public in April, 1963. In a little less than two hundred years after our first woman botanist had completed it, Jane Colden's manuscript came home.

Her special plant, however, had never gone away. In the cold cedar woods of the upper parts of North America, in Maine and New England and New York, in Canada and Michigan, Wisconsin, and Minnesota, Jane's flower, the little boreal goldthread, still blossoms every spring.

The period just preceding and during the remainder of the century following Jane Colden's death were years of discovery of plants in the eastern half of America. It was a time in which the European scientists, collectors, and wealthy men began to seek new and rare plants. They and their associates were discovering treasure of which they had never dreamed. An immense curiosity grew up about the plants and animals of North America, of South America, of the Pacific Islands, of every unknown place in the world. For the samplings which were coming to Europe were tantalizing. If such numbers of new species and remarkable plant forms could be found chiefly by untaught travelers and ignorant seamen who had simply picked them up as curiosities, then what might not a trained collector find?

It had been because of a sailor's fondness for his mother that the first fuchsia had come to England. In South America the young man had seen this extraordinarily beautiful plant with its ear-drop flowers of scarlet and purple, and had managed to obtain and bring home on the long voyage a plant for his mother. When a botanist, passing her cottage, saw this remarkable plant, he wanted to buy it immediately, but the woman was indignant and would not consider parting with it; the gift was not for sale. When the desperate botanist proposed to rent it for a month, she at last worriedly permitted him to take her treasure. During that time he successfully made cuttings from the first fuchsia in England, then returned the original plant to its owner. By such casual means very beautiful and unusual flowers were reaching Europe; it seemed a pity to have only such meager gleanings when intelligent collecting could scour a region for its botanical gems.

Thus it was the eager men of science and the generous men of wealth who made it possible for young men, often poor in pocket, to go out to the wildernesses of the world and, working on a modest salary, somewhat in the manner of a scientific grant today, to bring back plants for the cabinets and botanical gardens of Europe. When the Royal Horticultural Society of England was formed, a regular program of collecting and exploration for the benefit of Kew Gardens was carried on for many years.

It was a wonderful arrangement by which the wealthy persons, often too stout, too gouty, and too satisfied with comfort to endure the hardships of exploration, could finance the young, willing, energetic men who didn't worry about the undoubted discomfort and danger to which they were to be subjected for the sake of finding new species. Collecting plants became as much a hobby of the rich as was the collecting of paintings or postage stamps. There was an exciting challenge to urge one's personal collectors to find something that would surpass what the rival Lord Thus-

and-So's own collectors had discovered. It was often a risky business—for the plant collectors, that is. A good many of them did
not live to return, but that only lent the excitement of vicarious
danger to the stay-at-home gentlemen who were financing the
expeditions.

Peter Collinson, the London merchant who had been so much
interested in Jane Colden's career and in that of John and William
Bartram, encouraged Mark Catesby in London to study botany
and go to America to collect. Out of a small inheritance, Catesby

Collinsonia

at twenty-nine financed his own way at first and went out to
Virginia to visit his sister, who had married a doctor at Williamsburg. Mark Catesby stayed for seven years, and in that time he
discovered a portion of the natural world of America. Few plants
and birds had English names at that time; few had been identified
at all or classified in any way other than by the most casual common names. Catesby himself eventually had to create names for
many of the specimens he found—names which remain intact or
but little changed today from that time of wonder when nearly
everything in America was new and remarkable and to be discovered.

Catesby collected plants, shells, bird skins, birds' eggs, and anything else he could carry away from the delightful forests and
swamps and mountains lying behind the plantations of the Tidewater. There was a reason behind the collecting. Before he had
left England, some of his friends, especially Samuel Dale and
Peter Collinson, had made him promise faithfully to send back to

them pressed specimens, seeds, and plants, as well as other curiosities for their cabinets. The curios were easy enough to obtain and ship. The challenge was to transport living plants from America to England, for the long sea voyage was detrimental and usually fatal to green leaves and living roots. Dale and Collinson, insisting upon the live specimens from Virginia, instructed Catesby to plant young trees and shrubs in large tubs of earth. He must also see to it that fresh water for them was provided aboard ship because salt water would soon destroy them; he must pay a sailor to tend the tubs, protect them from excessive sunshine or wind, and carry them under cover during a storm.

The careful attention paid to the traveling plants succeeded. Dale and Collinson were delighted with what Catesby sent to them, but, although the specimens were impressive and exciting, his friends were greedy. They felt that, when his long visit to America was over, he might have made considerably better use

Catesby's Turkey Oak

of his stay. Looking back on the lazy days, the leisurely ramblings along the James, the York, and the Pamunkey rivers, poling an Indian dugout into cypress swamps simply for the adventure of it, attending house parties around the Tidewater, being entertained and enjoying the life at Williamsburg—he could see that he had indeed wasted much of his splendid opportunity.

No sooner was he at home again, with Dale and Collinson probing him for information he hadn't had time to gather, than he could see that he must go back. He would paint pictures of the natural productions of America, would write a book in which they might be featured. It had not been done before. Thus, if they had his guidebook for assistance, men who came after him might not have as hard a time in knowing what they found.

This was the time for someone with money to come to the aid of Mark Catesby and sponsor his quest and career. He had evidently gone through what funds he had had and could neither finance another journey to America nor publish his book when it was completed. The first step was to find provision for his lone expedition. Urged by Collinson and Dale, certain wealthy men with botanical interests subscribed to financing Mark Catesby's return to America in 1722, when he was thirty-nine years old. With funds enough so that he need not live with relatives or visit friends in order to eat or sleep, he ventured into new country, down into South Carolina and Florida, up the rivers to the wildernesses of the Smoky Mountains and the Shenandoahs. His discoveries, his adventures, and his paintings were rewarding.*

He had firmly expected to illustrate only plants in his book. When the birds and other creatures, however, always managed to insinuate themselves into his view, he was compelled to add them, perhaps out of desperation because of their insistence for attention. Consequently, the birds and flowers in his excellent paintings were often not in the right proportion to each other. Usually the flowers were too large for the size of the creatures sharing the page with them, creatures that had obviously been added as an afterthought.

His first sight of a bison in the Piedmont resulted in a charmingly detailed painting of that animal, highly animated and dramatic but wholly out of proportion in relation to the vegetation included in the plate. Perhaps to economize on paper, he placed it, very much dwarfed, on the same sheet as his painting of the black locust tree and a branch of the ivory-colored flowers. His well-colored depictions of botanical specimens were masterful; they were created with a care and skill which was rare for that day. Most botanical illustrators were still influenced largely by the stylized drawings and woodcuts of the herbals, but not Mark

* See Chapter 3, *Men, Birds and Adventure*, by Virginia S. Eifert.

Catesby. He drew from life and he painted to retain the look of life.

In 1724 Catesby returned to England with his paintings and his plants and set to work to create his great books, *The Natural History of Carolina, Florida, and the Bahama Islands: Containing the Figures of Birds, Beasts, Fishes, Serpents, Insects, and Plants: Particularly the Forest-Trees, Shrubs, and other Plants, not hitherto described or very incorrectly figured by Authors.* Since his former sponsors seemed unlikely to finance production of the two

Catesby's Southern Red Lily

luxurious and costly volumes, Catesby set about to do it all himself. When he had no more money, he learned how to make copperplates so that he could do the engraving himself; he hand-colored the finished prints. When he set out to find backers to subscribe to the books, he at first had little luck. Good Peter Collinson then lent him the money, without interest, and was in no hurry to be paid back. Someday when Catesby's books were selling well, then the debt could be repaid.

By 1746 the two splendid volumes were completed. Less than three years later, as if he had stayed alive only to complete the work, Mark Catesby was dead.

His were the first color-illustrated books of American botany and ornithology. In the long search for knowledge of American wildlife, men of science and collectors of fine books bought Catesby's books as standard reference works. Upon Catesby's discoveries (to which he gave names created in the old style, very long and descriptive, all in Latin) others, like the Bartrams and the

Coldens, based many of their own identifications. Linnaeus himself retained some of the Catesby names as part of the standard nomenclature.

At the same time that Catesby had been working, Johann Friedrich Gronovius in Holland, using plants sent to him from America, was compiling a book called *Flora Virginica*. Among his chief correspondents and collectors was the eminent John Clayton; he was one of the important yet little-known early American botan-

Spring Beauty—Claytonia

ists. In his honor Linnaeus named the spring beauty *Claytonia virginica*, and the interrupted fern *Osmunda Claytoniana*.

Interest in American plants grew among Europeans as men like Catesby, Gronovius and Linnaeus published catalogues and books of New World botany. The collectors in America were laying a solid foundation for American botany which has been added to and built upon ever since. They were still groping in a strange new world; they were making many pardonable mistakes; they were, however, all part of the great interlocking pattern of men and minds and natural history—joined for a time by Miss Jane Colden, who learned more about American plants than any other woman in America had ever troubled herself to learn before.

Mountain Laurel—Kalmia

5. Peter Kalm and the Mountain Laurel

He departed for America, his journal stated, "in the name of the Lord right after dinner," on a mission for the Swedish Academy of Sciences. The journey would result in an assortment of adventures ranging from shipwreck to Indians, from poison ivy to mosquitoes, and from a new wife to a collection of new and beautiful plants. He was Pehr Kalm, or Peter Kalm, as it usually is written, a young Swedish clergyman and botanist, a friend and fellow plantsman of Linnaeus, with whom he had gone on plant-collecting expeditions into Russia and the Ukraine in 1744. Possessing youth, physical stamina, diplomatic tact, and botanical knowledge, Kalm seemed to be the right man for the mission for which the academy chose him. Botanical collecting was not a career for the weak, the irresolute, or the timorous. One needed to know how to get along with Indians, how to subsist far from civilization, and how to collect, mount, and identify plants in wildernesses of the sternest sort.

Kalm had been sent to America to evaluate the useful vegeta-

tion which might be induced to grow in Sweden and thus augment that nation's sometimes insufficient natural resources.
Europe's famines were periodic and terrible, and thoughtful gov-
ernments now began to strive to find means of averting them by
introducing more kinds of food plants, finding reliable fodder
crops for animals, and widening and enriching the flora of the
countries of Europe. In studying and testing useful plants and
trees, the Swedish Academy of Sciences was comparable to the
modern agricultural experiment station.

Sweden was the first European nation to send a botanist to
scout the American landscape. By the time the other countries
tardily got around to sending their own collectors for the same
purpose, Peter Kalm had skimmed the cream of new varieties of
the eastern part of the country. To him alone Linnaeus credited
the discovery of at least sixty new species. Some suspicion was
expressed (possibly by rival collectors) that Kalm had lifted for
his own credit some of the plants which John Bartram had discovered, but, if he did, Bartram did not appear to mind, for they
remained good and fast friends.

Kalm set off with an expert gardener, Lars Jüngstrom, as his
companion and servant. After a journey involving a considerable
amount of trouble and delay at sea, including a shipwreck off the
coast of Norway, he and the sturdy Lars finally reached Philadelphia in September, 1748. Kalm brought letters of introduction
from friends in London, including Peter Collinson and Dr. Mitchell, to present to Benjamin Franklin and John Bartram. Franklin
immediately took the Swedish botanist under his care and attention, introducing him to his friends, and arranging for him to
borrow books without charge from the newly formed library in
Philadelphia. In his genial, expansive way, Franklin saw to it
that Kalm did what he wanted to do, met those whom he wanted
to see, and went where he wished to go.

On his very first day in America, the visitor was taken on a
sightseeing tour by two newly made Swedish friends, Bengston
and Hesselius. And on that day, Peter Kalm realized what he was
going to be up against among the wonderful confusion of the
American flora. He thought he had been prepared, in a measure,

for this new wealth of wilderness, but all his previous examination of herbarium specimens had not conditioned him for it. Even around Philadelphia he was both astonished and confused. He wrote:

Whenever I looked on the ground, I found everywhere such plants as I had never seen before. When I saw a tree I was forced to stop and ask its name of my companions. The first plant which struck my eyes was an *Andropogon,* or a kind of grass, and grass is a part of botany I always delighted in. I was seized with a great uneasiness at the thought of learning so many new and unknown parts of natural history.

His new friends would not let him rest. Three days after he had landed in Philadelphia, he was on his way with Jüngstrom and Gustave Hesselius southward on the highway to visit Mr. Bartram on his farm, about four miles out of Philadelphia. He

Persimmon

was eager to meet this man about whom he had heard so much that was good. Besides, en route, he had his first opportunity to really look at the landscape and examine some of the trees. On both sides of the road a great deciduous forest stretched as far as he could see. He found mulberry trees, oaks, walnuts, chestnuts, beeches, and sassafras; wild grape vines as big in diameter as his arm climbed up the trees or caused fences to sag with ripe fruit. He was tremendously impressed with the size and luxuriance of the vines. They embodied the strength, virility, and marrow of America. When Hesselius pointed out some persimmon trees across a marshy field, the men dismounted and walked over to examine them. Kalm admired the still-firm, pink-orange fruits, of which he wrote:

Its little apples looked very well, but are not fit for eating before the frost has affected them and then they have a fine taste. Mr. Hesselius gathered some of them and desired my servant to taste of this fruit of the land, but the poor credulous fellow had hardly bit into them when he felt the qualities they have before the frost has touched them, for they contracted his mouth so that he could hardly speak and got a very disagreeable taste. This disgusted him so much that he was with difficulty persuaded to taste of it during the whole of our stay in America, notwithstanding it loses all its acidity and acquires an agreeable flavor in autumn and towards the beginning of winter. For the fellow always imagined that though he should eat of them ever so late in the year they would still retain the same obnoxious taste.

Kalm set about immediately compiling a list of the trees which he found growing around Philadelphia. With fifty-eight species on the list, he wondered uneasily how many he had missed. He was out every day to examine woods and meadows; he roamed down byways and walked along the river banks. He collected as many seeds as he could find before winter set in, so that they might be sent at once while fresh to Sweden. Jüngstrom helped him.

Poor Lars Jüngstrom seemed to be the one to whom all misadventures happened. Among other things, he had been cautioned about poisonous plants. Nevertheless, when he and Kalm had handled both poison sumac and poison ivy, that first autumn, in preparing material to be sent to Sweden, he had scoffed at the warnings. Jüngstrom had been admonished about so many things: skunks . . . persimmons . . . Indian turnip. . . . The local people evidently enjoyed regaling the two Swedes with stories about such glorious American wonders as hoop snakes and horn snakes, eye-bungers, soap-boilers, and assorted hexes, so that Jüngstrom, in particular, had the feeling that he was being hoodwinked. Thus he stoically rejected everything he heard as mere myth or fable, if not downright lies.

He didn't believe in poison ivy or in poison sumac. He had proved it all a lie by apparently being immune to both. "Jüngstrom," said Kalm, "being a kind of philosopher in his own way, took nothing for granted by which he had no sufficient proofs, especially since he had had his own experience in the summer of the year 1748 to support the contrary opinion."

But the next summer, Jüngstrom's opinion was altered drastically. His hands were blistered; his skin itched; his eyes were swollen shut; he was miserable for many days. After that, whenever he came near either the ivy or the sumac, he was violently affected.

In cold scientific experimentation, Kalm tried to induce the poison to come on his own skin, but he decided at last that he

Poison Ivy

must be truly immune. He had spread poison ivy juice on his hands; he had rubbed the cut and broken branches on his hands; he had smelled of the leaves and carried juicy sections of bark around with him. He had certainly proved his immunity. Then, one hot summer day, he cut a branch of poison sumac. That night Kalm, the immune, awoke with a terrible itching on his eyelids. By morning, a breaking-out had burst on hands and face; he was terribly blistered and for a week was in great discomfort. Immunity, Kalm and Jüngstrom decided, was still another great American fable.

Altogether, with both good experiences and bad, Kalm and Jüngstrom put in a busy autumn. They assembled their collections and sent them to England before winter halted shipping. They would be forwarded from there to Sweden the following spring. That winter Kalm substituted as pastor in a small Reformed church in Raccoon, New Jersey, a Swedish settlement, where he evidently fell in love with the young widow of the recently deceased pastor whose place he was filling. The widow was both attractive and pleasant and was an excellent cook. Kalm dined often at her house. Before he returned to Sweden in 1751, he married the widow and took her home with him.

He enjoyed the interim with the little congregation of his own people. On Christmas Day he was happy to see the church decorated in his honor with boughs of the evergreen-leaved mountain laurel, for this, most of all, was his special plant. He had not been the one to discover it; the laurel had been sent to Europe at some previous time, but Linnaeus had chosen to name it *Kalmia* in honor of his botanical friend. Peter Kalm had not yet seen his namesake in bloom; he looked forward to that time, for this was considered to be one of America's most beautiful flowering shrubs. He had indeed been highly honored.

Pleasant as the winter was, he was impatient for spring. He and Jüngstrom had laid their plans for a trip to Canada, via the Hudson River and the wilderness of the North. It would be their most perilous journey, making their excursions around Pennsylvania and New Jersey seem very tame. The North was still a hazardous place to visit. The peace of Louisburg in Nova Scotia had been achieved only several years before, and it was known that a good many hostile Indians in the upper country still hungered for blood. Yet, dangerous or not, the sturdy botanist was compelled to explore for plants in Canada. Its climate was very much like that of Sweden and, it was believed, plants which throve in one place might do well in the other. Indians and difficulties notwithstanding, he would go. After a sedentary winter and the widow's excellent cooking, it was time he exerted himself on the mission for which he had been originally sent.

Fortunately for botanical pursuits, the keelboat on which Kalm and Jüngstrom embarked on an early summer day to travel up the Hudson moved neither very steadily nor very fast. It was frequently and providentially delayed when the wind was contrary. When this occurred and the craft had to tie up along the bank, the first ashore were the two from Sweden. Here they collected as fast as possible before the boatman's horn called them back and the vessel again proceeded slowly up the river. At Albany

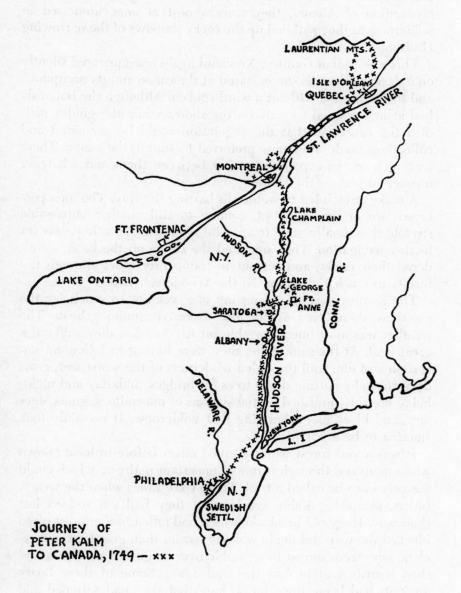

LAURENTIAN MTS.

ISLE D'ORLEANS

QUEBEC

ST. LAWRENCE RIVER

MONTREAL

LAKE
CHAMPLAIN

FT. FRONTENAC

HUDSON
R.

N.Y.

CONN.
R.

LAKE ONTARIO

LAKE
GEORGE

FT.
ANNE

SARATOGA →

ALBANY →

HUDSON RIVER

DELAWARE R.

NEW YORK

L. I.

PHILADELPHIA

N. J

SWEDISH
SETTL.

JOURNEY OF
PETER KALM
TO CANADA, 1749 — xxx

they acquired a canoe and two guides. Leaving the comparative civilization of Albany, they were almost at once immersed in wilderness as they paddled up the rocky shallows of the narrowing Hudson.

This was Indian country. Now and again one appeared silently on a dark, wilderness shore, stared at the canoe and its occupants, and then vanished without a word spoken. Although the botanists had at first wanted to walk on the shores while the guides paddled the canoe, so that the vegetation could be examined and collections made, they now preferred to stay in the canoe. There was, at least, an expanse of water between them and whatever menace lurked on the shore.

A rocky stretch led to waterfalls barring the way. The men portaged two of the falls, but, coming to still another impassable stretch, they finally gave it up. The river was quite hopeless for further navigation. They upended the canoe on the bank, shouldered their packs, and set out to tramp fifty miles through the forests to reach Fort Anne on the Woodcreek River.

The journey was an exhausting one, yet not so much for the exertions themselves as for the excessive summer heat. The weather was most uncomfortable for the Swedes; they suffered a great deal. At the same time, they were having to follow an uncertain and dim trail through a wilderness of the worst sort, cross deep rivers by cutting down trees for bridges, and, day and night, being horribly punished by the swarms of mosquitoes, gnats, deer flies, and black flies inhabiting that wilderness. It certainly took stamina to be a botanist.

It was a vast forest which seemed never before to have known white men, and through it ran an uncertain pathway which could scarcely even be called a trail. There were times when the weary, bitten, perspiring Kalm wondered if they had not indeed lost their way. Huge old hemlocks which had fallen in the forest often blocked the way and made walking harder than ever for the travelers, who were forced to scramble over or go around them, and then wearily had to find the trail again. Some of these hoary ancients had been down for so long that they had softened and decayed as long low ridges of red-brown, rotten, wet wood pulp which was thickly blanketed with mosses and other ground-cover plants. Solid though the hummocks appeared, they gave way disconcertingly when stepped upon.

Nights spent in the forest were not only much plagued by the insect hordes and the wood ticks, but also with a great dread of snakes and a fear of Indians. The men tried not to mention the Indians. Nonetheless, they thought of them, each in his private meditations. No Indians had been seen lately, but this did not mean that the savages had not seen the travelers or were not stealthily keeping watch over all their movements day and night. An uneasy feeling of being observed by unseen eyes was always with them.

The whole situation was a trial. To cap it, there were very few plants to be found in so old and shadowed a forest. The resilient earth was thickly covered with soft old fallen leaves or needles from innumerable autumns and with vast green beds of lush mosses that mantled everything, climbing up around the bases of the trees and studded now and again with ferns or ground pines. There were the few small plants characteristic of such a dark habitat—the club mosses and goldthread, twinflower, partridge berry, bunchberry, and Clintonia, but very little more.

The trees themselves were mainly hemlock, one of the darkest trees in America, huge and gloomy and quiet in its own perpetual shadow, where scarcely a bird raised its voice. Yet, in spite of the forest silence, the men found it difficult to sleep because now and again they were startled by a horrid wrenching, smashing, and thundering noise as another tree fell down. There was no wind, no storm. The trees simply might choose to fall of their own great weight and age when the time was right; the men could only hope that none deposited itself where the defenseless travelers lay.

At Fort Anne the men found no one. They had half expected to find at least a small garrison, perhaps some supplies, some shelter and protection, but the place was deserted by all but the mice. They were the kind which had no doubt journeyed in the soldiers' supplies, and when the fighting was over and the soldiers were gone away again, the creatures had remained as sole residents. In a realm of mice and men, it had not been the men who had survived.

At the mouse-haunted fort, Kalm, Jüngstrom and the two guides stayed for several days to rest and build a new canoe for the next lap of the journey. It was a relief to leave the oppressive closeness and menace of the forest and travel on the water again. The banks at first were low and open, then grew high and rocky, and the water was clear, transparent, and rapid. Hermit thrushes and white-throated sparrows sang in the undergrowth. Now and again beaver trails were visible along the banks. So vast was the lonely wilderness that it was with a sense of shock that the men one day found the remains of a campfire, could see from the way the grass was matted that a number of people had no doubt stayed there a little while before. There were still hot embers under gray ashes, but there was no indication of who the campers had been or where they had gone.

When, toward nightfall, the four met a French sergeant and five French soldiers traveling as protection with several Englishmen from Fort St. Frédéric to Saratoga, the botanists learned that six French Indians even then were out for blood in that very neighborhood. They had vowed revenge on the English for killing the brother of one of the tribesmen. Although the peace had indeed been declared, these Indians would have none of it until they themselves were personally avenged, and were on their way even now to the English settlements to accomplish their purpose.

Kalm, Jüngstrom, and the guides were somewhat shaken. They had slept in the same forest with a band of murderous Indians; they could have been waylaid and slaughtered; had, in fact, no doubt seen the very remains of the Indians' fire and their trampled trail, all without thought of their dreadful significance. Although the Indians were said to be intent on killing only an Englishman, it was a well-known fact that neither Iroquois nor Abenaki were inclined to distinguish between English or French. Whites were all one to these Indians, were indiscriminately classed as intruders, villains, and murderers, and were all fair game, Swedish botanists no doubt included.

For added safety that night, the travelers stayed together. The southbound group tried to persuade Kalm to change his mind and go back with them. It would be far safer, they said fearfully. But Kalm, trusting to the Almighty, insisted on continuing to Québec and to Montreal. This was his mission. He gave the canoe to the southbound party. The river above was impassable now for a long distance, for the French had felled trees across it during the recent war to prevent the English from coming up. But, the French soldiers assured him, they themselves had left their own birch canoes above this same barrier, and Kalm was welcome to take one in exchange. The Swedish adventurers thankfully reached Fort St. Frédéric at Crown Point at eight o'clock in the morning on July 2, 1750, where they were politely received by M. Lusignan, the governor of the fort, and were given breakfast and accommodations for that night.

Fort St. John, Kalm's next stop, lay at the top of Lake Champlain. By using a sail on the canoe, he and Jüngstrom made good time up the lake, and then proceeded on their way on foot—the guides were no longer with them, having fulfilled their contract by having taken them to Fort St. Frédéric—and reached Montreal at last.

After the harrowing trip northward through the great forests, it was especially gratifying and a splendid change when Kalm and his servant were greeted as celebrities in Montreal. Letters had come to the governor from officials in Sweden, telling him that an eminent Swedish scientist was on his way there to study the vegetation of Canada, and that it would be much appreciated if Canadians offered M. Kalm full hospitality. Charles Le Moyne de Longueil, the governor-general of Québec, sent word for Kalm to be permitted to travel at the expense of the King of France throughout French-Canada. Peter and Lars were given an elegant place of lodging; their meals were provided. After all the uneasiness, hardship, and discomfort of the forest, the attention and comfort, if a little overpowering at times, were both flattering and gratifying.

Kindly Dr. Gaultier was Kalm's guide and host during his stay. Since the doctor also was a botanist, his companionship was pleasant and useful—he knew where the best flowers and trees grew and could take Kalm there without wasted time. Also, if the visitor wished to examine farms and churches, the doctor took him there; even the private areas in the convents were opened to him. Such hospitality was astonishing. It made possible a highly profitable visit, and it may have been because of the doctor's kindness and co-operation in the cause of botany that Linnaeus named in his honor the charming northern plant, the wintergreen, *Gaultheria procumbens.*

When the Swedish visitors left Montreal, they traveled by boat down the St. Lawrence to Québec, making a pause to collect at Trois Rivières. Québec was impressive. After Montreal, which had had something of the aspect of a frontier town, Québec was like a French city on its high gray rock above the widening expanse of the St. Lawrence. From here Kalm went to the Île d'Orléans, to Lorette, to Montmorency Falls, and up into the foothills of the Laurentian Mountains. A Christianized Indian boy from the mission at Lorette was his guide.

In all his apparently aimless wandering, questioning, examining, collecting, and visiting, Peter Kalm had been learning something which had been suspected but which had not been proved. Botanists had noticed it here and there. Kalm and Linnaeus had conjectured about the matter, but they had not had a wide enough area of comparison, at that time, to be sure of the theory. He now verified that, going northward, an increasing number of Canadian plants were also native at a similar latitude in Sweden. On the north side of Québec, up into the Laurentians, at least one fourth of all the plants he came upon were identical with species he had seen in Sweden. Reindeer lichens were abundant in both places. As he ascended the mountains, nearly half of the plants in the sphagnum bogs and coniferous forests were similar to those he knew so well in the beloved coniferous woods and bogs of Scandinavia, all the way into Lapland itself. He was getting close to the secret of circumpolar botany. Later botanists would finish unfolding the story.

Kalm spent a fascinating summer and autumn. He collected seeds and pressed many plants, and in October he and Jüngstrom, somewhat reluctant to go so soon, started back for Philadelphia.

PETER KALM AND THE MOUNTAIN LAUREL ✤ 75

Canada was so large that Peter felt he had seen but little of it. He had been treated so well, besides, that now it was with a shock of disillusionment that, on expressing the wish to return by way of Fort Frontenac at the entrance to Lake Ontario, he was bluntly refused permission to do so. Kalm was hurt. He could not understand the rebuff. Heretofore, M. de Longueil had been so kind. The governor had known that the Swedish government itself had sent Kalm to Canada on a scientific mission, not a political one, so that whatever it was which the governor did not wish him to see at Fort Frontenac would be quite lost on a botanist who was intent only on flowers and trees.

Even after Kalm had written a long, pleading, explanatory letter in labored French to De Longueil, the governor still would neither relent nor explain why the Sieur Kalm was not permitted to pass by way of Fort Frontenac. Obviously, the French had an excellent reason for the refusal. He began now to wish that he knew what it was. There was nothing else to do, however, but to go back by way of the menacing forests and the execrable mosquitoes to the glittering Hudson which was, after all, no doubt the quickest route back into the safer and more amiable jurisdiction of the English.

Loaded with precious specimens and with innumerable envelopes of seeds, small packets that were neatly labeled and fastened securely to contain the hopeful embryos of Swedish gardens, he began his southward journey with the long-suffering Jüngstrom.

They were back in Philadelphia in November and had a meeting at once with John Bartram. Kalm had collected various seeds and specimens for him, and the latter wanted to know all the details of plant life in Canada. Peter wanted to discuss an additional fact which he felt he had proved, one which Bartram himself had proposed—that the farther north a species grew, one which was normally native farther south, the smaller it became. He had noticed, also, that a species which was most common southward quite often did not appear to have time to make its seeds if it was grown far northward where the summers, though possessed of longer days, had a shorter span before winter set in again. Although neither man knew about light-hours in relation to growth of plants, they had just come upon one of its obvious evidences.

Peter Kalm with his bride, and Lars Jüngstrom with the specimens, left America in 1751. Kalm hurried to confer with Linnaeus, who was waiting to add the list of new American species to his great work *Species Plantarum,* which was nearly ready for the printer. Linnaeus, immensely curious about what Kalm had found, would not complete his book without the addition of this incomparably rich new flora.

Kalm left his mark on the botany of America. The American plant manuals of today, as was Linnaeus's of yesterday, are studded with his name as discoverer of many species. To him was thus credited the Kalm's lobelia, a grass called *Bromus Kalmii,* and *Hypericum Kalmianum,* the Kalm's St. John's wort. His loveliest monument, however, is his namesake, the mountain laurel, *Kalmia latifolia,* blossoming in great clusters of crisp pink and white flower cups marked with carmine, in the mountains of the eastern states; in the smaller sheep laurel blooming in vivid pink; and, in the northern bogs, the dancing, wine-pink blossoms of the boreal bog laurel, *Kalmia polifolia,* all flowering in his honor every year in May and June.

Kalm's St. John's Wort

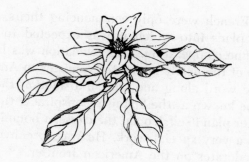

Large-leaved Magnolia

6. Michaux, the Botanist, and the Spanish Conspiracy

It was 1793. The guillotine had taken the heads of Louis XIV and a good many of his associates, and before the year was over his queen would also have been beheaded, together with Madame Roland and many of the Girondists themselves. Men no longer had titles. They were all to be called Citizen, and there were supposed to be only *liberté, égalité,* and *fraternité* abroad in France. In the violence of revolution and governmental upheaval, and in the fervor of success, the Girondists wished to convert the world to their own conception of a republic. France, in a murderous mood, while its inner throes were still causing blood to run in the streets, intended immediately to declare war on both its ancient enemies, England and Spain. Included in the careful details of this plan of conquest was America itself, which, assisting France at a certain point, was expected to have a vital part to play for the glory of the new France.

Not that America, shortly emerged from its own revolution at

77

the time when the French were only commencing theirs, knew anything about the plans into which it was expected to enter happily and with a fine co-operation. This revelation was largely left to the discretion of the newly appointed minister to America, who was to pass the word along and start into motion the plot which was later to be known as the Spanish Conspiracy; this was a branch of the master plan itself. To aid the cause, a botanist had been selected to do a very special work. He was to convey the information to confederates on the American frontier.

The minister who, with a fine diplomacy and dispatch, was to carry out the plans in America was the very young, earnest, handsome, red-haired Édmond Charles Édouard Genêt, called Citizen Genêt in the new mode of fraternizing. He, together with his associates in France, thought he understood the Americans. Give them plenty of whiskey at the conference table, it was agreed, and the Americans would assent to anything. They had indeed accomplished a very successful revolution which had thrown off the yoke of England, but, Genêt and the others felt, the whole thing might have had a cleaner impact if the slash of Madame Guillotine had been employed; she had a way of leaving no loose ends unaccounted for, left no discontented Tories to make trouble.

The principal reason for Genêt's coming to America was to start a war. Diplomatic mission or not, this was the real reason why the Girondists had sent Genêt and why the botanist had been appointed to play his own part in the plot. For France painfully resented the way in which its possessions in America had been taken over by the Spanish and the English—the one in New Orleans and in all of Louisiana; the other in Canada which, as history attested, had first been acquired by the French. The Spanish just now were blockading the mouth of the Mississippi River so that the Americans themselves did not have free access to their own river. Since the greatest amount of commerce and trade west of the Allegheny Mountains was carried to the markets of the South by means of the Ohio and the Mississippi rivers, the Spanish occupation of New Orleans and the lower Mississippi became more and more a source of irritation, anger, and frustration, particularly among the Kentuckians who had settled along the rivers earlier than some other people living west of the mountains. They depended for livelihood upon a large river trade.

Genêt, counting on this anger of the Americans at the presence of the Spanish on this river of the American flatboatmen, expected to gather recruits, money, and troops in America, then move down the Mississippi to attack and take New Orleans. At the same time, he would send privateers up the Mississippi delta from the Gulf of Mexico to attack and take New Orleans. It would be quick, relatively painless, and Spain would be neatly ousted forever from the Mississippi Valley. When this *coup* was accomplished, then the French and the Americans, thoroughly united now as brothers, would march on Québec and in the name of France take Canada.

Key persons and sympathizers in the conspiracy ranged from Philadelphia to St. Louis, from South Carolina to Kentucky. From them Genêt immediately began to recruit men and money. The whole affair, of course, despite the fact that it was known by so many people, had to be kept very quiet. There was little fear among the French that the Americans would object to the plot when it was revealed; the fear lay in the fact that entirely too many Americans had contact with the Spanish in New Orleans. Genêt could not risk releasing the news too soon.

Young, impetuous, impatient Citizen Genêt, in spite of his fine manners and his French persuasiveness, made no headway with President Washington. The latter was not in the mood to invite a war after he had only so recently and thankfully brought to a finish one which had been costly both in men and money. He had little patience with Genêt or with his outrageous Spanish Con- spiracy, and even when the young man pointed out that the Presi- dent had a debt of gratitude to pay to France for its help during the American Revolution, Washington would have nothing to do with the plan. Finally, while Washington's cold blue eyes bored into him, the furious young man quite forgot himself and actually ordered the President of the United States to call a special session of Congress which would at once openly declare war on the Spanish.

If there was anything the weary President of the United States wanted less in that final decade of the harried and momentous eighteenth century, it was another war. He was not to be pushed into it by any man, much less an importunate young upstart from the blood-dripping French Republic who had been making a nuisance of himself, not only with Washington, but with Jefferson and Madison and every other high official as well. Of them all, Jefferson had listened, and had let Genêt talk, had even given him the impression that he was actually in favor of the move, thus letting the young man reveal everything and display his full plans.

When help was ungratefully withheld by the American government, Genêt went into action on his own. If he could not have the backing of the American officials, he would succeed solely on behalf of France. There were many loyal French and many dedicated French sympathizers in America. They would all help.

Washington, realizing that Genêt was not going to be stopped, ordered the tumble-down old Illinois fort, Massac, on the Ohio, to be rebuilt and garrisoned. Meanwhile, down the Ohio River came flatboats loaded with ammunition which ostensibly was intended for defending the settlers from the Indians, but which actually was bound for an eventual attack on New Orleans.

But what of the botanist? Even before Genêt had left France with his orders for America, he had been told to select a botanist to travel west and pass the word of the conspiracy to key persons located out there. When someone recalled that a noted French botanist was already in America—André Michaux, who had gone there several years before to collect new trees for France—it was agreed that he would be perfect for the mission. Michaux was a mild-mannered man whom no one would suspect of having Girondist leanings or of being connected in any way with a plot. A simple man of science, his head down to observe plants, or his eyes looking upward to study the trees, he would be excellent as a middleman to carry letters of information to persons along the way.

André Michaux was born in Versailles in 1746, and because his father worked on a farm owned by the king at Versailles, the boy was brought up with the love of plants as part of his everyday life, and eventually became superintendent of the farm. He had had a classical education, a good one for the times, and had always geared himself and his attention to agriculture and to his great love, the wild trees and flowers of other places.

Life had seemed to be so simple, so destined, and so probable. He would be educated, would become superintendent of the king's farms, would marry, raise a family, and would bring up his own sons, if he had any, to succeed him on the farm, as his father had done. But life is seldom so simple, so peaceful, or so well ordered. When, in giving birth to their son, his lovely young wife, Cecile, died in 1770, Michaux's world fell into chaos.

Yet André Michaux had within himself a source of comfort which those who are naturalists will readily understand. In his grief he turned more fully to nature, and it was nature which, from that time onward, occupied his life with a reviving, deepening, and dedicated interest. He sought to identify every plant he found; when he had collected all there were to be had in his own area around Versailles, he traveled elsewhere. He went to England, to the Pyrenees, to the Auverge; he constantly studied plants and filled his life with them.

As soon as his son François André could walk and talk, Michaux took him on his excursions around Versailles and taught him very early the names of plants and birds. Then Michaux, leaving the boy in the care of relatives, went to Persia to collect more plants for the king's gardens. When he returned after his first experience in Oriental botany, homesick to see his child again, he expected to go back shortly to the East. This time he would take with him young François André, who was now fifteen and as enthralled with plants as was his father. But Michaux found that France had other plans for him.

It was 1785. France wished to introduce to its gardens and estates, chiefly those of the king himself at Versailles, some of the beautiful trees and flowers and other variously useful plants of the New World, as Kalm had done for Sweden, and Catesby and Bartram for England. Such countries were undeniably the richer for the additions, and France wished such riches, too, and did not, moreover, care to be outdone by other nations, even in botany. France, therefore, sent André Michaux to America to study forest trees. He was ordered to collect seeds and plants, experiment with those which he considered to be worthy of exporting, and send the satisfactory specimens to France. He was to acquire some land in America to use for an experimental botanical garden where his discoveries could be tested and propagated.

Michaux did not particularly wish to go to America. The Orient fascinated him far more. Nevertheless, in the autumn of 1785 he took the boy, who was excited by the adventure and eager to see America, and in obedience was on his way.

For eighteen months the two lived in New York. Whenever the weather permitted, the father and son ranged out to look for plants. It was not, however, Michaux decided, the best of all places for a plant collector to work or for him to maintain a botanical garden. He had heard that the Carolinas, having none of the severity of winter which, in the North, so shortened the growing season, were much kinder. Michaux and the boy moved to Charleston, bought some land about ten miles outside the city, and acquired lodgings in town. To the rich earth of the South he brought

Pink Azalea

the plants and trees which he hoped one day would grow in France. When enough specimens were ready to send, they were skillfully packed and shipped out at once; he was gratified to learn, after a long time had elapsed, that every one of them had not only survived the voyage but had set roots in the soil of France.

He and François André journeyed through the southern Appalachians and worked their way through the swampy wilderness of Florida. They sometimes hired an Indian guide, sometimes went alone, but always they were most fascinated with those southern mountains where so very much seemed to hide. In the high, misty ranges and valleys, Michaux felt, all of the most beautiful vegetation of eastern North America was concentrated. Here were glorious magnolias to thrill Europeans. There were splendid oaks, willows, gums, dogwoods, Amelanchiers, hawthorns, hollies, silver-bell trees, and so very many more. Every cove and peak revealed others. Yet, as the mountains drew him back again and again, he at the same time entertained thoughts of going west. A

Fairy Primrose

virtually unknown, plant-filled land lay beyond the Appalachians. No one as yet could even be sure how far beyond the mountains and the Mississippi River lay the Pacific Ocean. And no botanist had as yet gone far into that unknown wilderness to discover what plants grew there.

Instead of heading west, however, he undertook a long, difficult journey north to Hudson Bay and to the Lake Mistassini country to collect the small flowers of the tundra. One of these was that lovely little miniature, the pink and white fairy primrose. He found it on the stony tundra where gray lichens were interspersed with small blossoms forming a mosaic of color as far as he could see. An incredible, bleak landscape was accented with ragged spires of black spruce and with pools of cold melt-water and old snow left over from the spring's thawing. Growing out of crannies in the rocks, rooting in the meager, gritty soil, he found hosts of the tiny primroses. Each wiry little two-inch stalk was topped with the notch-petaled, lavender-pink, golden-eyed, fragrant blossoms. Although they shook incessantly in the cold wind, their fragility was a deceptive thing. Michaux saw with respect that any flower which could survive the punishment of the Lake Mistassini country had need to be exceedingly tough and inordinately tenacious of life.

He took specimens of this new species which he named *Primula mistassinica,* in honor of the chill, remote tundra world just south of Hudson Bay.

When he returned from the North, he found to his horror that François André had suffered an accident in which the sight of one eye had been destroyed. The father, conscience-stricken at having left the boy alone so long, sent him back to France to the care of relatives where he would complete his education. Since the youth wanted to be a botanist he must be well grounded in a classical education which included both Latin and Greek. François André thus went to France just as the French Revolution was beginning to alter forever the old way of life.

Meanwhile, left alone in America, André Michaux continued to yearn for the woods and watercourses of the western country. He must now go there to satisfy this yearning. Some time before he had become so imbued with this plan, however, Thomas Jefferson himself had had the compelling urge to send someone out to explore the land west of the Mississippi River. For many years the river had been the boundary, the wall, the barrier to the unknown. Because for a time the West was Spanish territory, the United States government needed to be very careful and very diplomatic about sending anyone on a mission across the river. Nevertheless, for years Jefferson had been desperate to know what was out there in a land which, he must have realized, one day would all become a part of the America he loved. He needed to know about the Indians who were rumored to be fierce and unreconciled to white men; he must quickly find out about the fur trade and just what the British were up to in that line on the West Coast. He wanted to know about all the plants, animals, and topographical details of the land lying vastly between the Mississippi River and the Pacific Ocean.

Although several expeditions had been promoted by Jefferson, they had all failed to materialize. Now in January, 1793, Thomas Jefferson, Vice-President of the United States, and André Michaux, botanist from France, sat together in a quiet room in the Vice-President's house in Philadelphia to discuss still another expedition to survey that mysterious western country.

The botanist, a superb woodsman and walker, free to go anywhere, possessing the intelligence to know what he was seeing and to evaluate it, could think of nothing he would rather do than make that scientific expedition west for Jefferson. The American Philosophical Society had been prevailed upon to take up a subscription to finance this project. The money, besides, would be welcome to Michaux, since he had been low in funds for some time. Ever since the revolution had changed matters in France, and the king, who had originally employed him, had lost his head, the French government had owed the botanist much money in back wages.

Michaux happily accepted Jefferson's proposal. He was to start up the Missouri River in the spring and continue as far as he could, hopefully, to the Rocky Mountains, find his way over the mountains to the Columbia River, and thence go to the Pacific Ocean. In proposing and in accepting, neither man realized the huge extent of the journey involved nor the dangers likely to be incurred by a lone naturalist. Happily ignorant of the true state of things, Michaux went to lay his plans; he expected to depart as soon as the ice went out of the rivers in April.

Suddenly, on April 10, 1793, Citizen Édmond Charles Genêt sent word to M. André Michaux to wait upon him on an urgent matter of great importance to them all, especially to France.

Astonished to be so summoned, André went at once. He had thought for a long time that France had forgotten him. After a perfunctory greeting, Genêt set about at once to explain the situation, and as the botanist listened, he became more and more dumbfounded and then very much excited. A faithful Frenchman, a loyal Girondist, even though he had been in America when the French Revolution had occurred and the Girondists had surged into power, he fully believed in the reasons for revolution, in the abolishment of monarchy, and in the rights of the common man. News of the guillotine's work had gratified, not horrified him. Actually he had been astonished at the reaction of the Americans to this concise form of punishment.

Now as the minister, being certain that the doors were closed and that no one listened at the windows where a gentle breeze wafted the curtains, explained very carefully what the plans were to be, Michaux realized that he had been selected to serve France in an even greater manner than that of finding new trees and flowers. He was to convey the word of a great conspiracy which would bring about an uprising and the overthrow of the Spanish in America. To free the mouth of the Mississippi from the Spanish would greatly assist both France and America in future plans. It was a most worthy cause.

The plan as related by Genêt sounded eminently sensible, yet a nagging question lay in Michaux's mind—just how will the Americans fall in with a thing like this? He had lived for too long in America not to know something of the people and about how they reacted to various situations. He expressed none of his inner doubts to the minister, who had not asked his opinion. At the same time, Genêt had no way of knowing that, even as he was giving the botanist his orders, matters in France were changing, and the orders which he himself had been given by the Girondists were now largely canceled because of new changes in party and power. Even without the consent and sponsorship of France, Genêt was the sort of man who would pursue his own war. He wrote, boasting:

"I excite the Canadians . . . I arm the Kentuckians . . . I prepare a naval expedition [to take] New Orleans."

Genêt gave Michaux a packet of letters addressed to influential men along the way, men living from Pittsburgh to Kentucky. It was in Louisville, Kentucky, that he would contact one of the most important figures in the conspiracy, George Rogers Clark.

Clark was fully in the mood to work with France rather than with the United States. Bitter and disillusioned with the way his country had treated him—he who with his own money had financed the project to free the West, and who had never been reimbursed; he who had fought in the Indian wars and had paid his men out of his own pocket; he who had been eminently loyal and heartbreakingly unappreciated—he was finished with the United States. Clark was ready to secure land outside of that nation's authority, start his own community, his own state, his own way of life. He wrote to Genêt in February:

I can raise abundance of men in this Western country—men as well Americans as French. . . . Out of Kentucky, Cumberland, the settlement on Holston, the Wabash and the Illinois I can (by my name alone) raise fifteen hundred brave men, or thereabouts—and the French at St. Louis and throughout the rest of Louisiana, together with the American Spanish subjects at the Natchez, would, I am sure of it (for they all know me) flock to my standards. . . . With the first fifteen hundred alone I can take the whole of Louisiana for France.

Clark even promised that, if the mouth of the Mississippi were sufficiently well blocked by French frigates, he could take New Orleans with ease, as well as Pensacola, and might even go west to Santa Fé and Mexico, saying, "if France will be hearty and secret in this business, my success borders on certainty."

It was not so much the secrecy of the French which George Rogers Clark needed, however. He, Washington, Jefferson, and Genêt were all doing business with a spy who, as a general in the United States Army, was fooling them all and selling the information he obtained to the Spanish at New Orleans. James Wilkinson, associate of Clark, knew what was going on on both sides and he turned that knowledge to his full advantage.

Something was wrong. Clark was aware of it, but he could not put his finger on it. He had a feeling of uneasiness even before Michaux finally reached him with Genêt's latest directions.

It was July 15, 1793, before André Michaux could take leave of Citizen Genêt with his final orders, his letters of introduction, and his collecting case and plant presses. He departed unobtrusively on horseback at ten o'clock at night to avoid the oppressive heat in Philadelphia which had been so unbearable that summer. In the city, the heat, the flies, the stench, the growing number of deaths during an epidemic of the plague, all had been a horror to endure. The heat had abated somewhat with the coming of night over the steaming city; a hot summer moon rose full as Michaux and two companions rode out of Philadelphia. The two were M. Humeau and M. LeBlanc, both agents of the minister and ardent republicans.

LeBlanc had been promised to be made mayor of New Orleans when that city should fall to the French. The two were sent along as far as Pittsburgh to act as friendly companions of the botanist, but it was suspected by Michaux that they were sent as a precaution to make certain he was well started on his way and did not deviate from his course.

It was a relief, as they headed into the mountains, for the men to get away at last from the fetid, overpowering atmosphere of heat and death in the plague-ridden city, and to travel into the cool, wonderfully clean air of the high country. Michaux, finding all familiar things here, botanized along the way. He was no doubt too late to find the wild American strawberries of which he was so fond, and which he and the young François André had discovered several years before—sweet, richly flavored, exquisite wild fruits which were so much more delicious than any they had known in France. He had sent back plants which, when crossed with European species, produced improved strawberries. Later on, they came back to be planted in gardens in America. Now, plodding up the steep grades of the mountain roads, he was interested in all that he found alongside. If there were no strawberries, then perhaps there was something else. His letters and his mission may have weighed heavily in his saddlebags, but his plant presses were still light, and his mind was always on what he might find to add to them.

Pittsburgh. He did not like Pittsburgh, but was compelled to linger here while he waited for a boat to take him down the Ohio. The river was very low, too low for the flatboats to navigate. Waiting for a rise, they were all tied up along the waterfront of the city, while the botanist, who could never waste precious time, roamed the river bottomlands and hills.

August 12. It was at last raining in the mountains, and two days

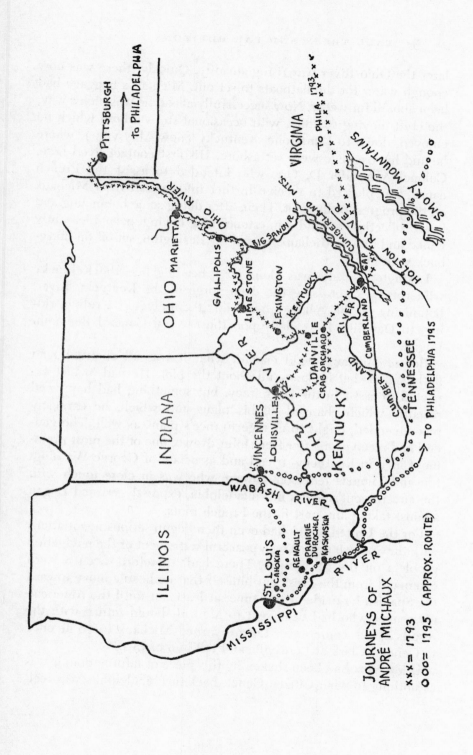

JOURNEYS OF
ANDRÉ MICHAUX

xxx = 1793
ooo = 1795 (APPROX. ROUTE)

later the Ohio River was rising steadily. Quickly, there was now enough water for the flatboats to set out. Michaux's baggage had been aboard for weeks. Now successfully afloat, he was on his way, the craft moving steadily, with occasional stops ashore which he enjoyed, down to Limestone, Kentucky (now Maysville), where he and his baggage were set ashore. His first contact lived here. Colonel Alexander D. Orr, who intended to assist the French cause, was pleased to receive further information from Michaux on the progress of the plot. Then, after doing some botanizing and fossil-collecting around Limestone, and having been pleasantly wined and dined, Michaux, bound for Lexington, set off on horseback with Orr.

Lexington . . . he was alone now, heading into the Kentucky wilderness . . . crossing the deep gorge of the Kentucky River, botanizing on the yellow-brown cliffs themselves . . . riding, riding, to Danville. He called upon the revered General Benjamin Logan.

Before Michaux could explain what it was all about, Logan flatly stated that he knew all about the plot. He had in fact expected to assist in the conspiracy, but something had happened recently which changed all his plans and which, he earnestly recommended, ought to change France's plans as well. The general had received a letter from John Brown, one of the most prominent men in Kentucky, friend and associate of George Washington and Thomas Jefferson. Brown, who was in close touch with the government's affairs in Philadelphia, expressly warned Logan against taking any part in the French plots.

For the United States had even then begun serious negotiations with the Spanish regarding a peaceful settlement of the navigation problem on the Mississippi. There had, therefore, been a stern command from the President himself that no hostile move toward the Spanish be made at this time, at least not until the American messenger who had been sent to Madrid should return with the results of the conference. Logan warned Michaux to go at once to General Clark at Louisville and tell him of this.

If Michaux had been shaken by this piece of information, it was as nothing to what Citizen Genêt, back in Philadelphia, was feeling after the letter which he had just received from his superiors in France. During that abominable summer, Genêt had been laboring mightily in the heat and stench and flies, and after all his efforts he had received what was perhaps the most irritating communication from France that any minister could have had. The Republic had changed, M. Deforgues had written, and called him a fool, and worse, to thus endanger the delicate status of relations between America and France. M. Deforgues not only gave Genêt definite orders to desist, but coldly informed the young minister that a replacement was even now on his way to relieve him of his duties. It was as if Genêt had deliberately set himself to cause trouble between the two countries, yet he had in fact only been doing what he had originally been ordered to do when he left France in 1792. The whole problem was the unstable French government. One party in power had told him to do thus and so; immediately, another party canceled everything the other had done. Working at such a great distance from France as he was, with a resultant slowness of the mails, it was no wonder Genêt could not keep up with what was happening.

The petulant minister, stung by the accusations of Deforgues, would not listen. He had gone too far to stop now. If no one else wanted a war, he would promote his own.

Although Michaux out in Kentucky did not know what was happening in the East, he did know that Logan's communication aroused in him a vague unease. Nevertheless, since he was acting for France and not for America, he continued on his way to deliver all of his letters. His duty finished in Danville, Michaux then did some delightful and relaxing botanizing in the wild country surrounding the curious, conical hills called the Knobs, where he found several totally unknown species which pleased him immensely. The new plants were worth all the frustrations and the lack of accomplishment on this oddly unsatisfactory mission.

It seemed to Michaux that the men to whom he was supposedly carrying highly secret news were the ones who instead were giving him news, and that what he heard was not reassuring. Nor did he feel better when he reached Louisville at last and sought out George Rogers Clark on his country estate. Clark's enthusiasm for the project, which Michaux had been led to expect, had become curiously diminished.

General Clark, of all people in Kentucky, had originally seemed to be most interested in the conspiracy. Now he appeared to be reluctant to commit himself. Events were brewing, he said cautiously, which opposed such a project at this time, and he did not consider that he should enter into it just now. For somehow, in some way, Clark sensed that the news had been leaking out. He evidently did not suspect the true villain, James Wilkinson, but he knew that someone was informing the Spanish about what was going on. Clark's spies had told him that the Spaniards were preparing a defense of their own in New Orleans, for they had evidently received word of the rumor that the Americans intended sending an army down the Mississippi. Clark was uneasy. He would not continue with anything further on the project at this time.

At Louisville, Michaux's work for France was done. If Clark wished to bow out, then let him. Now Michaux the botanist was free to continue with his herborizing. It was, after all, his primary and most-beloved work; he had not, he decided, been cut out for a spy. Although the journey apparently had been worth very little from the standpoint of politics, it had become, from the standpoint of plants, an increasing challenge and delight. If he had not been able to carry out those cherished plans for Jefferson's western journey, Michaux had at least managed to come farther west than he had ever been before, and perhaps farther than any other botanist had ever ventured and collected.

Wishing he had the opportunity to see more of Kentucky while he was there, he nevertheless managed to cover a different route on his return to Philadelphia. Going by way of the often mountainous Wilderness Road and collecting specimens of flowers, trees, and seeds all the way, he went part of the way with an armed escort. This pioneer route was still rugged and unsafe for a lone man. He prudently waited at Crab Orchard until enough people had assembled to traverse the country of the Shawnee and Cherokee. Twelve men, armed and mounted, were sufficient; they were ready for one hundred and thirty miles of wilderness, Indians

—and, for Michaux, of totally delightful trees and flowers that grew between Crab Orchard and the next station, the Holston settlement in Tennessee.

It was mid-November, and the weather had grown chill. Yet, going south through the Cumberland Valley, with the mountains high and purple and difficult of ascent, he found many plants which were so well protected from the cold and the increasing frosts that they appeared to be still as green as in summer. He paused long enough to examine lovely, wet, streamside banks covered with masses of the curious climbing ferns, *Lygodium.* Bearing small, lettuce-green, palmate leaves on tough, tendrilly, twining stems often three feet long, which were unable to stand alone, the climbing ferns had scrambled over dripping banks or up among the woody stalks of the rhododendron bushes and laurels to make a charming little jungle of themselves.

On this journey he may have seen once more one of his most delightful discoveries, the Catawba rhododendron which he and

Yellow Wood

his son had first come upon when its glowing lavender-pink blossoms ornamented the heights of the Smoky Mountains. And now, on a sleeting, disagreeable day in eastern Tennessee when he was returning to the East from a political mission which apparently held no great success for anyone, he found a new tree that was destined to be one of the rarest in America. The yellow wood, *Cladrastis lutea,* was one of his best finds. Although he did not see it in bloom until much later, he knew he had something fine— this graceful, locustlike tree with smooth gray bark and twisted boughs, and hanks of pendant, beanlike seeds still hanging from the twigs. He could not reach any of the seeds, but was kindly assisted in getting them by a soldier at the Holston fort who cut down a tree for him so that he might get at them.

On December 12, 1793, André Michaux reached Philadelphia with his plant presses full and his bags bulging with seeds and roots, but with little news, either good or bad, for Genêt. Whether or not he had been of any use to France, Michaux humbly did not know. But he did know that he had seen some beautiful country, that he had come upon an unrealized wealth of plants, and that he had not had enough time to collect as he would have wished. He longed to go back when spring was on the hills and he could see that landscape and its treasures all in bloom.

By the next year the Spanish Conspiracy was all over. Genêt was ousted and the ship which had brought his replacement, it was rumored, had also brought along a guillotine by which to dispose of Genêt in an official manner, since President George Washington, for all his irritation at the young man's inexcusable actions, seemed indisposed to give him up for proper execution in France. It had, therefore, been decided in France that America should have the opportunity to see first hand how political problems were handled elsewhere. It was recommended that the guillotine be accepted as a gift from the French government for future use in America. Washington, the story went, would not permit this ghastly gift to be brought ashore, much less used, nor would he give up poor Edmond Genêt, who had tried so very hard to cause a war. Now that the whole thing was over without undue incident and no bloodshed, Washington could forgive. Genêt himself had no wish to go back to France. He had fallen in love with Governor Clinton's daughter Cornelia (the Clintonia plant was named for the governor) and in 1794 they were married. He never returned to his homeland. America might have destroyed him politically, but it had given new life to Genêt as a man.

Michaux, meanwhile, in the spring of 1795 set out happily on another expedition to the West. He was not, however, going up the Missouri River to carry out that old plan of Jefferson's. As far as Michaux was concerned, that had been canceled. He would never see the Rocky Mountains or the Columbia River. No one would blaze that route to the Pacific Ocean until 1803, when Lewis and Clark would set out and would succeed at last in the project which Jefferson had planned for so long.

Nevertheless, on his own Michaux went as far west as the Mississippi River. He set out in the spring, worked his way slowly through all the flowering beauty of the Smoky Mountains and the Cumberlands, collected flame azaleas and several magnolias, and new oaks, and in hot summer weather finally came to Vincennes,

Michaux's Basket Oak

Indiana, on the Wabash. He hired a Shawnee Indian and his squaw as guide and cook on the route across the tall-grass prairies of Illinois. The Spartina and turkey-foot grasses were taller than a man riding a horse; he was awed by the height and apparently endless sweep of the grass, but it was not easy to traverse. In islands of forest marking with dark green the sunny prairie, he came upon the bur oak and the basket oak, both new to science; the latter became known as *Quercus Michauxii* in his honor.

He visited the old French villages of Kaskaskia, Cahokia, Prairie du Rocher, and Renault along the Mississippi and paused in St. Louis for a few days before heading southeast again. An autumn journey was rewarding, for it provided the great harvest of seeds that were ripe for the gathering. Seeds were easy to transport; a great many could be carried in a small space. Concise, compressed, exciting, they contained in themselves the beauty of gardens and forests as yet unplanted and still to be realized. He would carry them to France, where a portion of America would grow and blossom, and he would remember, seeing them there, what it had been like to roam the wilderness of that far land.

Michaux was one of the most noted botanical visitors to America. He was a foreigner, as Catesby, Kalm, Pursh, and Douglas were foreigners, but he and they left a portion of themselves as part of the heritage of America. It lies along the trails they traveled and in the plants which they discovered and named.

Although he had been instrumental in finding and naming so many new plants, it had happened more than once, on his journeyings, that Michaux had come upon flowers which were so strange to him that he could not find a name or even decide on the genus. He left these problems for later leisure when he had returned to France and had the time and the facilities for study. One plant, however, he never identified. He had found it in his beloved southern mountains. It had low, rounded, scalloped, evergreen leaves rising around the base, and several stalks bearing white bells. The leaves were somewhat like those of the Galax, but the flowers—he had never seen anything like those purple-marked white bells with notched petals, could not even begin to guess at what they were. Intending to work on them later, he pressed several good specimens. They were in his collections when he at last returned to France.

He could not collect forever without any money. The French Revolution was over; changes in government had altered his own status. For the past seven years he had received none of his pay as botanist to the king. Since there was no longer a king, Michaux was presumably no longer on government pay, except for what Genêt had given him. Therefore, it was time to go back and remedy his finances, as well as to get reacquainted with his beloved son, who was now a man. Michaux packed his dried and mounted plants, crated many living trees, shrubs, wild flowers, and ferns. He was taking back with him a large part of America, yet he felt that there had been, in a measure, a fair exchange. Wherever he had gone, he had carried packets of seeds from France, and had left these with cabin folk in the mountains, or with people with whom he visited in the towns. In taking back with him some of the most beautiful trees and wild flowers in

America, he was also leaving some of the beauty of the Old World. Now he was anxious to determine how all the plants which he had sent during the past years were growing in the lovely gardens at Versailles.

It was a long and tempestuous journey. At last when his ship was nearing the coast of Holland, it encountered still another storm. The vessel fought valiantly and lost. It crashed on the off-shore rocks. When he knew that the ship was breaking up, Michaux calmly tied himself to a plank. With his herbarium lashed securely with him, he waited for whatever would happen. Then as the ship creaked and groaned and he could hear it coming apart, the spikes wrenching out, the waves pounding the wreck, masts falling, the men yelling, he suddenly knew a great surge and found himself and his board washed into the furious sea. Losing consciousness amid the buffeting of the waves, he knew nothing more until he opened his eyes and found himself cast upon a beach. He was still lashed to the plank, but his precious bundles of plants were gone. Groaning, he lifted his head weakly, saw a fire, saw people, and his head fell back again. His plants . . . lost. . . .

Someone saw the movement and ran to look at him—he had been thought dead. The men, Hollanders all, hastened to untie the wet ropes and help him to the fire to warm himself. They gave him something to eat and drink. When he could speak, he asked about his herbarium. At first they did not know what he was talking about, but then one saw the sodden bundles lying where they had been tossed to one side, and brought them to him. Michaux felt a surge of relief as great as the wave which had thrown him off the sinking ship. His souvenirs of America were safe. Everything else he had owned was gone, all but the herbarium, and his life. These two were enough.

The botanist was taken to the nearest town and given lodging. While he recovered from the ordeal, he worked to save his precious specimens from ruin. Soaked, soggy, salt-encrusted, they were in a terrible state. Carefully removing the plants from their moldy papers, he washed them thoroughly in buckets of fresh water. Painstakingly laying out and drying the rejuvenated specimens from the forests and meadows of America, he remounted them on fresh papers until they were almost as good as before. One of those which he saved may have been the unknown plant with the Galaxlike leaves which had been discovered in the southern mountains.

Somehow, no one ever took the trouble to classify that unknown specimen. Michaux, not long after the shipwreck, reached France. After seeing with grief that most of the trees and flowers he had sent from America had been destroyed by the revolution, he was soon afterward sent to Africa to collect. Only a short time later he perished of a fever on the island of Madagascar. His collections were put away in the archives of the Jardins des Plantes in Paris, where for a time they were all but forgotten.

François Michaux, following his father's example, went to America, traveled widely as a botanist, and collected even more extensively than his parent. He apparently did not examine in detail the latter's collections, and therefore must have missed the unknown plant. Evidently, also, neither he nor anyone else ever came upon it growing in the wild.

In 1839, a botanist once more bent his attention upon the unknown flowers. Young Dr. Asa Gray of New York, visiting Europe for the first time and making a reverent pilgrimage to the centers of botanical study in the Old World, came to Paris and visited the famous Jardins des Plantes. Calling for the plant collections of André Michaux, he carefully leafed through all the sheets. Most of the specimens had been named, but he came at last upon a special paper containing a plant with Galaxlike leaves and a cluster of brownish bells which must have once been white. In spite of the soaking in the shipwreck, the plant seemed to be in good condition. There was no name on it.

Gray, an expert botanist, decided with mounting excitement that this was a new genus. What Michaux had found, he, Gray, would have the pleasure of giving a name and rescuing from oblivion. He thereupon christened it *Shortia* in honor of Dr. Charles W. Short, a botanist of note. Since the leaves were really very much like the glossy, evergreen foliage of the Galax, he added a specific name of *galacifolia*. He noted that Michaux had found it in the southern mountains, a broad and remote area containing many hidden coves and mountainsides where, like Bartram's lost Franklinia, anything might hide for centuries. And, like the Franklinia, the Shortia evidently had never been seen or collected by anyone since its discoverer added it to his collecting case.

In 1840, when Asa Gray made a collecting trip to the same mountains, he failed to find the coveted plant. Further exploration by other men did not reveal its whereabouts. Gray was beginning to feel that the Shortia was another one of the lost plants of America—as if the mysterious southern mountains were possessed of strange spirits which permitted plants to be briefly seen, then whisked them away again.

Thirty-eight years later, an herb collector in North Carolina came upon a strange plant with white bells and rounded, scalloped, leathery, evergreen leaves something like those of the Galax. The collector, who knew Gray, sent him a plant. When it blossomed in Gray's garden at Harvard, the aging botanist was so delighted with it that he knelt and caressed the charming flowers. The next year he and Mrs. Gray traveled down to North Carolina to find the Shortia for themselves, but they arrived too late to see it in flower. They did see the plants and became acquainted with

Shortia

the woods in which they grew—perhaps, indeed, they were the very woods in which Michaux himself had first found the lovely Shortia.

Gray, Michaux, and an herb collector, over a period of more than eighty-five years, had worked together to name and add one new, beautiful wild flower to the flora of North America. It was only one of the strange and wonderful links which bound naturalists and botanists in a vast network of knowledge, friendship, co-operation, persistence, and discovery. Whether they would be discovered through political intrigue, by the harsh realities of exploration, or by happy accident, there were thousands of plants which would demand that many men and many years be required before they were all found out, given names, their habitat known, and their range charted.

Columnar Coneflower

7. Nuttall

He was too deeply interested in plants and birds to be very much concerned with his own safety. The American wilderness and its natural productions were so magnificent compared with what he had known in England that Thomas Nuttall could endure almost anything—and often did—for the privilege of traveling and collecting in America. He was a shy and retiring man who did not fancy crowds of people, or even half a dozen of them, unless they were botanists or ornithologists who spoke his language. Few men knew him well or understood him. Those who did were men of his sort. Any oddities in which he indulged were comprehended at once by such people as the great botanists John Torrey, or François André Michaux, or John Townsend, or Lewis D. Schweinitz, the expert on the grasses and sedges. They understood Nuttall because they were all inwardly a little bit like him. They were dedicated to plants, but some of them were not so adventurous in spirit, nor so daring for discovery, nor so careless of public comment, as was the shy, odd naturalist, the soft-spoken Thomas Nuttall.

Because he had no family and was alone, all of wild America

was his. Withdrawn, ingrown, and aloof, Nuttall blossomed into a different person when he was in the woods, upon the lakes, or in the presence of the wild things. He was alike a botanist and an ornithologist. His books on both subjects, standard reference works during the first three quarters of the nineteenth century, revealed little of the real man or of the downright adventure which had made them possible.

For, to look at him, one might never have suspected that the blood of explorers like Lewis and Clark or the mountain men flowed in those veins, nor that those cold, unsociable eyes in the plain, solemn face ever burned with the glory of seeing sights which no other white man and certainly no other botanist had ever beheld. Thomas Nuttall was both an enigma and a delight. One could always make a kindly joke about him; but one could also, in great scientific respect, await his finds and the publication of his lists of exciting new plants.

Botanists in the East, however, could at times only complain rather plaintively, yet tolerantly, that they did wish that old Nuttall were not quite so slow. He took all the time in the world—sometimes years—before he got around to publishing his discoveries. Until he did, other botanists felt constrained in publishing theirs, lest they transgress on botanical ethics in naming a plant already given a legal epithet by the eminent Thomas Nuttall.

At first, when Nuttall was but newly come to America, he was enthralled with the vegetation and birds of the East. In pursuit of them he made excursions out of Philadelphia, loading his vasculum and keeping his plant presses full on every journey and sitting up late to classify new finds. For, although the East had been very well supplied for many years with naturalists, both native and foreign, it was still possible to find new species. Nuttall, roaming like a meditative tortoise with his eyes to the ground, was finding species, both among birds and plants, which his predecessors and contemporaries had evidently missed. He roamed with a dedicated thoroughness along the Schuylkill and the Wissahickon, explored the fascinating Pine Barrens and the sphagnum bogs of New Jersey, tramped in the salt marshes of Long Island, and went to the wilds of southern Delaware. He ventured into the Great Cypress Swamp and climbed the Shenandoah Mountains, and he journeyed north to the Hudson River.

When, in 1809, he made a plant-collecting trip for Professor Benjamin Smith Barton, an eminent botanist, through Pennsylvania and New York to Niagara Falls and Lake Erie, he went on foot most of the way. On foot, by rowboat, or by canoe—these modes of travel were much more suited to a botanist's pursuits than public transportation, which went too speedily for his liking. Then in 1810, when he was twenty-four, it was in utmost pleasure and freedom that Thomas Nuttall walked around the southern shore of Lake Erie and reached the pioneer community of Detroit. There he purchased a canoe and, alone and apparently unconcerned with the danger of accident or the fact that he was going into the Indian country of the wilderness of the often tempestuous Great Lakes, he paddled off along the coastline of Lake Huron to Mackinac Island. It lay like a pine-masted, rocky ship at the entrance to Lake Michigan.

This was magnificent—this great, cold, deep-blue lake, and this rocky island capped with trees, sturdy against storms and waves, a fortress against the whole wild West. He roamed the hilly island, saw the natural bridge and, from the fort on the hill, looked meditatively out to the shimmering glitter of the Straits of Mackinac that led to the mysteries of Lake Michigan.

On the island at that time, outfitting and preparing for a long journey west, were the bearded, buck-skinned members of the Astoria expedition. Eventually going up the Missouri River after a winter's wait in St. Louis, they would proceed overland to the Columbia River from the headwaters of the Missouri, and then to the Pacific to bring back furs from Fort Astoria on the coast. Amused by the botanist's eagerness for new adventure, they offered to take him along, assuring him the protection of the expedition. This protection, everyone told Nuttall soberly, not joking, was an absolute necessity for anyone going into the country of the Sioux, the Blackfoot, and the Shoshone, whose domain was the western plains and mountains. Anyone going there was in great danger, but there was a greater measure of safety in numbers. It was certainly unwise for a lone man to go into that cruel country, as Nuttall in eagerness would have rashly attempted. Nuttall, feeling that he could survive anything and go anywhere if only there were new flowers and trees for him, eagerly accepted the invitation of the Astoria men.

However, since he wished to travel in as leisurely a manner as possible, he set off alone in his canoe, well ahead of the expedition, to rendezvous eventually in St. Louis.

It was early June. As he paddled along the north shore of upper Lake Michigan and passed down into Green Bay, he saw on the beach sand drifts and masses of small, lavender-blue irises. Two-inch, azure-and-gold blossoms standing above short stems, they were tiny things, exquisite in their miniature perfection. The yellow rhizomes were set in a shallow rooting in the sand. From the roots had grown broad beds of the slender little leaves which seemed to be everywhere on those upper beaches of Lake Michigan. He was delighted to have the privilege of naming a new and lovely blossom—the dwarf lake iris, *Iris lacustris*. He may also have found nearby the tiny, sturdy, fairy primrose on the rocks and sands of the lake shores; it was the same species which Michaux had discovered in the Lake Mistassini country below Hudson Bay. Nuttall saw beach grasses he had known in New Jersey, and the sand-holding beach peas which he had found on Long Island shores. He may have wondered at their presence here so far from salt water.

With many halts ashore to look at plants, Nuttall in his canoe came down through Green Bay to the old French town of that name, and went on down the Fox River to the westward. He portaged from the lower Fox to the Wisconsin River not far away, and headed down that stream as Jolliet and Marquette had done long before, to finally meet the Mississippi.

The Mississippi! A journey upon it was a revelatory experience for any man. The river took him on his way past noble cliffs and hills of its widening valley. The landscape, in fact, was all so fascinating that it was impossible to hurry, and he was thankful that he had no need to do so. Plant presses full, he was almost sorry when at last he came to the muddy waterfront at St. Louis. Then he forgot his regret in the pleasure of meeting the estimable botanist John Bradbury, who also was going on the Astoria expedition in the spring. For Nuttall to find a man of his own kind so far away from Philadelphia was highly gratifying. The two spent the remainder of the season in botanizing around St. Louis.

In May, when the ice had surged down past the yellow sand bars at the mouth of the Missouri, it was time to set off with the men of the Astoria expedition who, navigating their huge, heavy mackinaw boats, started slowly up the cantankerous Missouri. It was only five years since Lewis and Clark had come back from the Columbia by this same route, and it was still the same long, hard way. Like the men of the Lewis and Clark expedition, the Astoria men had to pole the boats or get out on shore and pull them against the current. It took the sheer force of man power and will power to move upstream boats like these. Nuttall took his turn at the pulling, or cordelling. He rather enjoyed it. It served to put him on the shore where the plants were.

When he was compelled to ride in one of the boats, he impatiently scanned the banks. It was downright painful, both to him and to Bradbury, to have to pass things they both wished to examine. When the boats did pull in to the shore, Nuttall was always the first man off, sometimes in his haste plunging knee-deep in thick mud. The French-Canadian boatmen often laughed at him among themselves. They pointed to their heads, describing circles with their fingers, and grinned knowingly, calling the Englishman "le fou," because of his odd ways and extraordinary enthusiasms over trifling and useless weeds and bushes.

He went as far as Fort Mandan. This was the place at which the entire Lewis and Clark expedition had had to spend the winter of 1804-05, for that expedition had traveled slowly. The Astoria men, with greater speed, would continue up the narrowing and often rocky Missouri to the mountains, would pack their gear over the high trails to the Columbia, and would proceed in canoes down the Columbia to Fort Astoria near the sea, all in one season. The next year they would come back with their rich bales of furs.

For a reason which Nuttall did not explain in his journal, he decided to go only as far as Fort Mandan in North Dakota, and then return later on a downbound trading boat. He may have wished for more time and leisure to collect in the area; the Astoria men were in too great a hurry to suit him. Besides, on the high plains of North Dakota almost everything he saw was new. He collected as fast as he could; there was so very much territory to cover, and he was fearful of missing something. The Mandan Indians, tolerating his odd ways, were kind. They must have remembered those other strange-acting white men a few years ago who also had been curiously excited over birds and plants which to the Mandan people were commonplace. For Nuttall was seeing some of the same plants and animals as had Lewis and Clark, and the thought no doubt pleased him. Besides, because that earlier expedition had moved at so much slower a pace and thus had not reached this location where they built Fort Mandan until late autumn, and had departed in the spring, they had not, therefore, seen the summer plants which Nuttall was now joyfully discovering.

Although the Mandans were his friends, some of the other local Indians were not always so understanding or so kind. An experience with the latter left him somewhat shaken. On a borrowed horse, he had ridden out over the great lift of the cloud-capped hills a number of miles from the fort, and had found himself rather unwisely out of communication with everyone. The prairie flowers were such a delight that he had not noticed Indians on horseback bearing down upon him. They were not Mandans. He did not know who they were, and at first was unconcerned, but when they commenced yelling rather rudely at him, brandishing their feathered spears, he suddenly felt a sickening jolt of fear. He kicked his horse in the ribs and the animal took off at a fast canter up the hills, down the great inclines on the other side, startling the western meadowlarks, and passing flowers which he glimpsed only briefly in his flight. For a time he could hear his pursuers behind him, then after a while, hearing nothing, he risked a quick glance over his shoulder. He appeared to be quite alone on the vast, rolling plain, but he continued nevertheless at a fast pace up the next slope, then rode down into a ravine which was overgrown with cottonwoods and buffalo-berry bushes along a small stream. His knees strangely rubbery, he got down off his horse and held the beast's muzzle closed so that, if the Indians were near, still looking for him, the animal would not betray him by whinnying. There was, however, no further sound of pursuit. As his heart began to beat more steadily again and he relaxed, he drank from the stream as his horse was doing, and wished he had something to eat. There was nothing here but the cottonwood leaves and twigs, delicacies which the animal seemed to relish. Nuttall remembered that one of the Astoria men had told him that the Indian horses liked cottonwood in preference to grass or oats. But he himself could not eat such fodder.

When he was rested and felt that the Indians had gone on after having had their fun with him, he mounted and rode cautiously up the slope. All about him the prairie hills, almost treeless except along the streams and ravines, stretched like a frozen undulation of green and brown and flowers. With no idea which way the fort lay, he had no notion of how far he had come. The sky had become overcast; the sun could not be used for a compass. He simply went on, just to be moving, and by evening was worn out, terribly hungry, and a little bit alarmed. This was a delightful country to explore, but he disliked having to sleep out here at the mercy of the wolves, Indians, and unknown dangers which were likely to be more dreadful than either.

By the next day he was even more thoroughly lost and a great deal more hungry. Then, as he was down in a ravine in search of wild currants, a jack rabbit, bouncing up suddenly under his horse's nose, startled the animal so that it took off at a fast run over the hills. As the man stood and watched the animal go, he felt very small, very defenseless, and very much alone in an immense, impersonal, and implacable landscape.

Hopeless, miserable, and famished, he lay down under a bush. The plants in his collecting case, all his new locoweeds and his penstemons and verbenas and composites, were withering for want of being pressed. But he did not know which way to go, and he was too tired. . . .

He woke suddenly. An Indian, staring in surprise at the man on the ground, stood over him. The Indian did not appear hostile, only very much astonished and concerned. By the broad and

placid face, Nuttall decided he was a Mandan and therefore sure
to be his friend. He sat up, gestured to convey the message of his
need. The Indian helped the white man up on the former's horse,
and the two rode slowly together back over the prairie hills, and
plodded down at last to where the sluggish, winding expanse of
the Missouri flowed past the little fort.

Before the river had frozen that autumn, Nuttall returned to
civilization. He rode by keelboat to St. Louis and then continued

Western Sand Lily Found by Nuttall

to New Orleans, landing there shortly before the great earthquake
of 1811 shook up and altered a large part of the Mississippi Valley.

From New Orleans he took passage on a ship bound for England. He had plants and seeds, mineral and rock specimens, shells
and birds' eggs, notes and material on Indians, and the skins of
birds and mammals. His reason for going to London, not to Philadelphia, was a product of the times, since politics must influence
even botanists. There were rumors of conflict and he wanted to
be at home before it broke. He stayed in England during the War
of 1812. Then, in 1815, when it was over and peace again lay between England and America, he came back. America called to
him and he could not stay away.

For the West still beckoned, the West of the upper reaches of
the Arkansas River, along whose length no naturalist had penetrated. Nuttall boldly determined to go there alone. Other men
only hindered his pursuits.

On October 2, 1818, with a minimum of money and equipment,

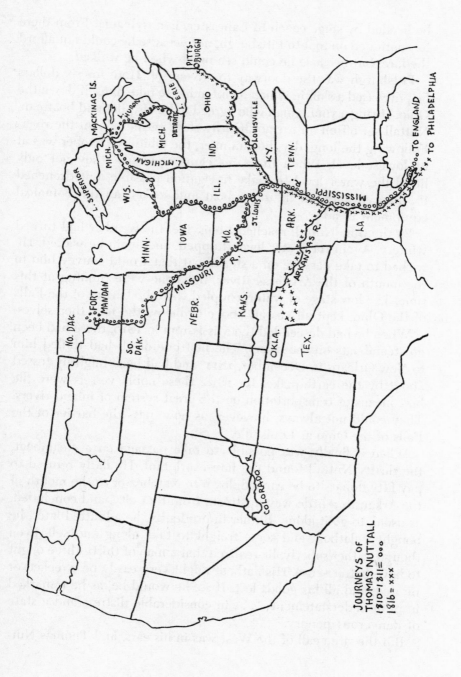

JOURNEYS OF
THOMAS NUTTALL
1810-1811 = ooo
1816 = xxx

he headed by stage coach to Lancaster, Pennsylvania. From there he continued on foot to Pittsburgh, not because he could not afford the fare, but because he could see more when he walked.

Pittsburgh was the doorway to adventure. Here for six dollars he purchased a skiff in which he intended to take himself down the Ohio. Hiring a young man who agreed to be his pilot and boatman, Nuttall set off on an extraordinarily slow journey down the river. Following the long, dry, hot summer, the Ohio by October was at its lowest level. When the rains came suddenly, they not only lifted the water level to make navigation possible, but drenched the two voyagers in the open boat and soaked all the botanical equipment besides.

By the time Nuttall reached Louisville, the weather had turned wintery. Again the water level dropped, and again he waited. He wished to take passage on a steamboat that would convey him to the mouth of the Arkansas River, but none were running at this time. The low stage of water, coupled with the barrier of the Falls of the Ohio, kept all vessels above or below the falls themselves.

When he had descended the Mississippi in 1810, there had been no steamboats on the rivers. The first one of all had trailed him to New Orleans in December, 1811, and had been mightily tossed about by the earthquake. By 1818, these noble vessels were the best means of transportation on the great system of inland rivers. They could not always, however, as now, pass the barrier of the Falls of the Ohio at Louisville.

When it finally was possible to take passage on a steamboat, the thrifty Nuttall found the fare exorbitant. He flatly refused to pay fifty dollars to be carried down to Natchez or to the mouth of the Arkansas a little way upstream from that city, and concluded, as usual, to go it alone. Rather imprudently, he admitted later, he bought a flatboat and some freight to take along and sell, even though, rather soberly, he realized that a man of thirty-three ought to have better sense. His flatboat might very easily be wrecked or pirated, and all his goods lost. If so, he would be, as he remarked in prim understatement: ". . . in considerable distress and a state of dangerous penury."

But the siren call of the West was in his ears, and Thomas Nut-

Above: Ocean shores, besieged by waves, devoid of plant life, greeted the firstcomers to America and are but little changed today. *Below:* The tundra is dark and apparently lifeless, yet, as Linnaeus discovered, it is the abode of low, bright flowers and multitudes of lichens which thrive in the harsh habitat above timberline.

Above left: The shooting star (Dodecatheon Meadia) was discovered and named in the eighteenth century by Mark Catesby. *Above right:* The Osage orange was first seen by Meriwether Lewis in about 1805, but was unnamed by him. Seeds brought to Philadelphia grew into trees which were named *Toxylon pomiferum* by Constantine Rafinesque. *Below left:* Mountain laurel was named *Kalmia latifolia* by Linnaeus in honor of the botanical explorer Peter Kalm, who came to America from Sweden in 1748. *Below right:* The great white magnolia (Magnolia grandiflora) charmed John and William Bartram when they ventured south to Florida in 1765.

The dark forests of America both lured and repelled the men who first explored them. Catesby, the Bartrams, the Coldens, Kalm, Michaux, and Nuttall discovered many new plants and animals in the American wilderness.

Above left: The forests along the Cumberland River in Kentucky are but little changed from the time when André Michaux followed this route down the valley to Cumberland Gap. *Above right:* The Catawba rhododendron was one of the discoveries of André Michaux and his son in the Smoky Mountains. *Below left:* The flame azalea (Rhododendron calendulaceum), whose flowers vary from rust-red and shades of orange to a glowing lemon-yellow, was one of the discoveries of Michaux in the southern mountains. *Below right:* The climbing fern (Lygodium palmatum) is still found along the valley of the Cumberland, especially on the west bank of the river at Cumberland Falls State Park.

Above left: The dwarf lake iris (Iris lacustris) was discovered and named by Thomas Nuttall on his canoe journey through upper Lake Michigan in 1809. *Above right:* White marsh marigolds (Caltha leptosepala) blossom in snow water and at the edges of old drifts on high mountains when spring is ascending the peaks. *Below left:* The wild hyacinth or camas (Camassia esculenta) blossoms in an expanse of pale lavender-blue that shimmers like lake water under the prairie sun. *Below right:* Mariposa lilies were one of the charming finds of Nuttall, Douglas, and Bigelow in the West. The species shown is *Calochortus Gunnisoni,* named for Captain J. W. Gunnison, who was killed by Indians in Utah.

The Mississippi River is a river of constant turbulence and change. Thomas Nuttall and his flatboat met eddies and rough water like this on the lower Mississippi when he navigated downstream in 1816.

Spring with its new leaves and its myriads of blossoms came early along the Arkansas when Nuttall set out to discover new plants along its upper reaches.

The great forests along Lake Superior in the nineteenth century stretched as far as men could see. Time has made many changes, yet today in the Porcupine Mountains of the upper peninsula of Michigan this forest appears much as it did a century and a half ago when the Cass expedition passed this way.

Above left: The bearberry (Arctostaphylos Uva-ursi) was one of the plants found by David Bates Douglass on the Cass expedition. *Above right:* Moss campion (Silene acaulis) grows in low mats on the rocks above timberline of high mountains and tundra. *Below:* Too shallow for a canoe, the narrowing Mississippi is overhung with willows and alders. Beyond this tangle lies the shining Lake Itasca, headwaters of the Mississippi.

Tall and exceedingly straight, the Douglas firs populate whole forests.

Above: At the edge of melting snow, David Douglas found small flowers just emerging from the stern winter of the high country. Here grew the charming glacier lily (Erythronium grandiflorum). *Below:* Even in the cruel, stern landscape of high mountains in the Tetons, Rockies, Cascades, Sierras, and Selkirks, which would seem devoid of life, small flowers bloom intrepidly in their brief and brilliant summertime.

Above left: The Joshua tree, often twenty to thirty feet high, is one of the few plants of any height in the desert. It was discovered by John Engelmann who named it *Yucca brevifolia. Above right:* Yuccas of many species were seen on western expeditions and continue to ornament arid places today. They are in the lily family. *Below:* Forbidding, arid, shimmering red-brown in the hot sunlight, the mesas of Utah, New Mexico, and Arizona were part of the landscape as expeditions worked their painful way to the coast.

The beauty of a fern which lived in a swamp in Illinois 250 million years ago is preserved in its perfection in a nodule of stone found in a strip mine. This is *Neuropteris gigantea*, from the Mazon Creek fossil beds a hundred miles southwest of Chicago. (Photo courtesy of Charles W. Hodge, from *Leaves and Stems from Fossil Forests*, by Raymond E. Janssen, Illinois State Museum, 1939.)

Above left: Thoreau and his companions canoed on lakes and streams of the Allagash wilderness, waters margined with spruce and hemlock, with marshy edges and deep forest. *Above right:* The purple-fringed orchis (Habenaria psychodes) was seen from the stage coach going to Moose-head Lake, and then was collected from a marshy shore in the wilderness. *Below left:* "It was a mossy swamp . . . ready to echo the growl of a bear, the howl of a wolf, or the scream of a panther. . . ." From Thoreau, *The Maine Woods. Below right:* Twinflower growing in a delightful carpet of pink bells has delighted naturalists from the time of Linnaeus to the present. Asa Gray, Sir Joseph Hooker, and John Muir found it for the first time in California.

The contrasting textures and forms of southern vegetation pleased John Muir on his thousand-mile walk to the Gulf. In Georgia and Florida he saw the stark fans of the palmettoes rise to meet the soft veils of Spanish moss hanging from live oaks and bald cypresses.

Tall and silent in their own world high in the mountains, the Sequoias are the climax of American vegetation. Not many miles away from their Sierra eyrie, timeless ocean waves break on sterile sands as they did when men first came to America to seek its tall trees and follow its far horizons.

tall, eminent naturalist and scientist, soon to take over a teaching position at Harvard University (if he survived this wild journey), found himself the owner of a weather-beaten old flatboat loaded with corn and salt pork which, under his inexpert command, was heading down the Ohio River for the Mississippi. He was not, after all, quite alone, and was rather glad about that. He had met at Louisville a gentleman he called Mr. G., and the latter's sixteen-year-old son Edwin, who were seeking transportation to New Orleans and who, like Nuttall, were feeling the pinch of finances when confronted with the steamboat fare. They went along with Nuttall to assist, rather ineptly most of the time, with the navigation as far as the mouth of the Arkansas.

They were a curiously innocent and improbable trio—two totally inexperienced men and a boy—to be starting forth to make their fortunes down the wild rivers. Nuttall, however, with enthusiasm that seldom flagged, felt he was ready for anything. He had purchased and studied a copy of Zadok Cramer's river guide, the *Navigator,* and with this he expected to be able to cope with any emergency which the rivers might offer.

The journey indeed began very well. Another rise in the Ohio had caused the water to rise ten feet in the night. The flatboat floated nobly over the falls and moved rapidly all night, so that within twenty-four hours they had gone nearly eighty miles without having had to exert any labor at all. It was an unimaginable luxury to be able to recline in the small cabin on the deck of the flatboat, snug with a hearth and table and their beds, and let the wintery river take them easily on their way.

The temperature, however, had dropped to ten degrees above zero, and Nuttall soberly hoped they could get far enough southward before the rivers froze. There already was ice in the backwaters and in the big cypress swamp behind old Fort Massac on the Illinois shore. They had had to break the ice away from the boat before they could start out that morning. Then, as the flatboat finally neared the point where the Ohio joined the flow of the Mississippi, the craft was suddenly surrounded and buffeted by crowding, rumbling, menacing masses of floating ice. Ice filled the Mississippi and the mouth of the Ohio itself. It was all unharnessed danger. Other downbound boats pulled over to the Kentucky shore and tied up. Nuttall and his companions prudently followed the example. They all stayed against the Kentucky banks for the rest of that day and the next, and fearfully watched the great brown-white ice floes crowding past, saw massive floating trees going by, and wondered what would happen to a defenseless flatboat caught up in it. The ice seemed to be traveling at great speed, while its supplies, coming down from upriver, seemed to be inexhaustible.

When, on the third day, the boats finally ventured out into the moving ice, the men found to their relief that traveling with it was not half so hazardous as they had expected it to be. The ice itself, in fact, was not so much the danger as were the half-sunken or submerged dead trees. Thrust into the mud and water, they presented sudden death to any vessel which rammed against them. In most cases, however, the flatboats themselves either crashed heavily into the unseen obstructions and then floated off again, or simply skimmed over the surface. Boats like these had no low-lying hulls to be impaled upon or ruined by snags.

The three spent a thoroughly dismal Christmas Day floating down the dreary middle of the ice-filled Mississippi. The huge, drifting trees, like battering rams, were still so much an ever-present terror, there was really no time to ponder upon the holiday, regret the situation, or feel any homesickness. Nuttall had never seen such awful things as the floating trees which came rushing along, plunging and shoving as if alive and determined to ram them. They were as powerful and uncontrollable as maddened buffalo; and they invariably seemed headed straight for the defenseless flatboat.

With his copy of the invaluable *Navigator* always open and at hand, Nuttall tried his best to follow the course of the channel as laid out by the expert Cramer. But this river changed so. He was certain that it had altered since Cramer had last passed that way, as it no doubt had. The Mississippi is an endlessly variable river; it is full of whims and moods; and today it is surely not as it was yesterday, nor as it will be tomorrow.

In some alarm, he read of immensely dangerous sandbars and

masses of trees at Plum Point Reach, obstructions which almost closed the course of the river. He read carefully the directions for going through, and consequently thought he was prepared. Yet, as the boat approached what he felt must be that vicinity, he saw nothing in the river to indicate any such danger. He relaxed. Cramer had no doubt been wrong. Then, next day, it all loomed before him—massive yellow sandbars half filling the broad river, and unimagined numbers of huge and scraggy dead trees thrown about as in a mighty barrier. Against this the current broke with a roar and the sound of a tempestuous millrace, while fresh debris from upstream was constantly being carried down and piled against the ugly barricade.

The three worked cautiously and fearfully to control the heavy boat on its headlong way down through the narrow space which remained open for a passage. Meanwhile, the rain, which had been spattering down dismally at intervals all that winter day, now descended in such a frightful deluge that visibility was cut to nothing, and the situation, if possible, was worsened. Hoping that things might improve next day, they tried to move in toward the shore and tie up for the night, but, in working in that direction, the vessel was caught in a great whirlpool which, with astonishing speed and power, was eating out a bank. With great effort they finally worked the boat out of the eddy and through the barricade, and to the shore. Then, quite worn out, they tied up to a willow about ten miles above the first Chickasaw Bluff in Tennessee. Next morning they awoke to a thick white fog which hid the river, the shores, and everything else within ten paces of where they stood.

When the fog finally lifted and they could see their way, they pushed cautiously into the river again. It was the last day of the year—it was New Year's Eve—when they came into a perfect forest of snags and sawyers. No one knew which way to steer around these obstacles, but the boat decided the matter for itself by striking with great force against a dead tree. Edwin was thrown headlong into the river and the steering oar went with him. Careening, the boat almost overturned; the barrels and boxes slid toward the low part, tilting it even more. Fortunately, the heavy boat righted itself with a splash, while the boy, who could swim, got within reach and was hauled aboard. The steering oar, however, was stuck in the mud, wedged among snags, and could not be dislodged.

With all the difficulties along the way, Nuttall had had little time or opportunity to get ashore to look for plants. When he did have the chance to do so, he hardly cared to try it again. All along the way down this interminable river, he had grown more and more disillusioned with the sameness of the vegetation. He had expected—what had he expected? Something different, at any rate. He knew the southeastern states and their plants and felt that the lower Mississippi ought to be a combination of both the South and the West, with a flora which would indeed be both rich and unique, to say the least. Instead, he was oppressed by the silence of the dank, dark, bottomland forests. They were of leafless soft maple, elm, sycamore, and willow; endless stretches of brown willows and cottonwoods margined the banks. Under them grew few plants. For hundreds of miles along the river, besides, there were no towns, no people, no houses. It was, in fact, very much the way it still is today. He exclaimed:

How many ages may yet elapse before these luxuriant wilds of the Mississippi can enumerate a population equal to the Tartarian deserts! At present all is irksome silence and gloomy solitude, such as to inspire the mind with horror.

The three travelers finally reached the mouth of the Arkansas River. Nuttall waited until his friends had found passage on a boat bound for New Orleans, then he turned into the Arkansas at last and worked his way six miles to the port of Ozark, where he disposed of his freight and sold his boat in the bargain.

Thomas Nuttall now was rather suddenly becoming excited. To be on the wild waters of this river at last, a river whose source no one at that time knew, a river coming out of the far western mountains and plains, leading him to who knew what new experiences, had brought him to the threshold of adventure.

Even the weather delighted him. It seemed suddenly far removed from the snow and ice and cold which he had left behind him on the ice-jammed, snag-bound, foggy Mississippi and the Ohio. On January 13, he noted that the temperature along the

Arkansas was as mild and warm as May, and he was downright happy to find yellow Senecio flowers on the river banks actually coming into bloom. Still, he was impatient with that persistently general sameness of the forests along the river. He wrote discontentedly:

No change, that I can remark, yet exists in the vegetation, and the scenery is almost destitute of everything which is agreeable to human nature; nothing yet appears but one vast trackless wilderness of trees, a dead solemnity, where the human voice is never heard to echo, where not even the ruins of the humblest kind recall its history to mind, or prove the past dominion of man. All is rude nature as it sprang into existence, still preserving its primeval type, its unreclaimed exuberance.

By the twenty-second of January, in a world filled with bird song, he had worked his way up the river by canoe to Arkansas Post. Cardinals saluted the spring; bluebirds caroled in the willows; mockingbirds chortled from treetops. The ground blossomed with drifts of white Arabis. The bird-foot violets bloomed, and he found the little greenish mousetail, *Myosurus*. Acres of bluets with gold centers laid a lovely azure hue over slopes and meadows. But these species, welcome as they were, had all been discovered and named by others, some by Michaux, some by Elliott. Nuttall longed for something new, something that was his, something that would cause a stir in botanical circles.

He had brought a letter of introduction to a Mr. Bougie at Arkansas Post, and having contacted him, was invited to stay and his baggage was thereupon brought up by a slave to the gentleman's house. It was while Thomas Nuttall was visiting here and botanizing that he began at last to find some of the new plants for which he longed. One of the first was a new wild onion with greenish flowers. It was not much, but it was a hopeful start.

When in late February he secured passage on a large skiff which was going up the Arkansas to Fort Smith, the peach trees, dogwood, and redbud were in bloom, and spring was everywhere. Upstream, he saw that the vegetation was changing perceptibly from the river bottom flora to something else, the Western look. Three hundred miles up the river, at the Cadron, when he came upon his first cactus, *Cactus ferox,* he felt with delight that it boded his nearness to the great Mexican desert.

He met a trapper named Lee and threw in his lot with him. Lee was an agreeable person, was going pretty much in the direction in which Nuttall wished, and moreover the man seemed to know the topography of the whole vast area in a way which could be invaluable to the botanist. Now the rewards for the long, tedious trip at last were coming to Nuttall. The whole early summer landscape was gloriously painted with color and fragrance and blossoms, and he was finding new species every day. There were large-flowered, richly colored, blue-purple spiderworts, a new Collinsia, various species of extraordinarily brilliant pink phlox and verbena, acres of the pale-blue wild hyacinth or camas, violet-blue Ixia and blue-eyed grass which were mingled with the yellow Senecio and golden puccoon, accented with the splendid wands of large, pea-shaped, cream-colored flowers of the first of the wild indigo.

Our route was continued through prairies, occasionally divided by sombre belts of timber, which serve to mark the source of the rivulets. These vast plains, beautiful almost as the fancied Elysium, were now enamelled with innumerable flowers, among the most splendid of which were the azure Larkspur, gilded Coreopsides, Rudbeckias, fragrant Phloxes, and the purple Psilotria. Serene and charming as the blissful regions of fancy, nothing here appeared to exist but which contributes to harmony. . . . Several large circumscribed tracts were perfectly gilded with millions of the flowers of Rudbeckia amplexicaulis, bordered by other irregular snow-white fields of a new species of Coriandrum.

If Thomas Nuttall let himself wonder at the drive which sent him off into wildernesses and into a variety of uncomfortable situations, he knew the answer lay in those new and beautiful blossoms and in the trees and landscapes of America such as he was exploring now.

Nuttall and the trapper worked their way inland from Fort Smith and explored widely across the country beyond, a countryside ranging from arid sand hills that were like the desert, to deep forests, and back again to the incredible, flower-filled prairie.

It may have seemed an Elysium, but this was because it was

early in the summer. With the change of the season to late summer, so also did Thomas Nuttall's physical situation change for the worse. Illness had begun to tarnish the glow of the experience. He had been troubled all summer with attacks of malaria, and, as he and Lee camped on August 14, north of the present city of Tulsa, Nuttall felt with foreboding the preliminary chills preceding another bout with the fever. Long, broiling periods of fever and delirium followed by enervating chills which shook him until he felt as if his very bones were turning to ice and quivering within him—it was a terrible punishment to endure, a poor way to pursue a botanizing trip, yet there was nothing to be done about it out here in the wilderness except to suffer and endure.

Since Lee's horse had torn itself on briers and had a bad wound on its back which bothered it a great deal and made it unable to carry a load, Thomas Nuttall felt he himself might take the time to be ill. While the horses rested and recovered he, too, might get through the worst of the coming bout. Since there were no trees and the heat ranged between ninety and a hundred degrees on those summer days, he managed to crawl beneath some bushes before he gave himself up to illness and lapsed into delirium. Lee, seeing that his friend was situated as comfortably as possible and being unable to do anything further for him, went out to hunt beaver. When he returned that evening, the botanist was much worse.

The heat was excessive. The dreadful blowflies, attracted by the meat brought to the camp, were unbearable. They swarmed stickily on everything that was alive or dead; they penetrated wounds on the horses and laid eggs even in the men's clothing, until everything was polluted. The flies were a nightmare. In an effort to escape them, the tormented horses were moved. But five miles down a creek it was really little better, and Nuttall, in being moved, had fainted twice on his horse and had nearly fallen off before his distracted companion caught him. On that day, Nuttall could have passed all the rarest flowers in the West and they would have meant nothing. It was all he could do to maintain sanity, equilibrium, and courage. There was only the terrible baking heat, the lack of shade in the glare, and the awful flies. Lee was worried.

He suggested that they return to the nearest settlement or fort before Nuttall grew too weak to be transported. He was certainly going to die out here.

In spite of his weakness, Nuttall could not bear to retreat without exploring any more of this compelling new Western country. Faintly, he begged Lee to wait two days longer, promising that then, if he were no better, he would consent to return, or else he would be dead.

Lee said nothing more about it. His horse could carry nothing, and they had no other way of getting themselves and their baggage back to civilization except to use the remaining horse twice—to double-trip until the injured one was better. The flies continued in abundance, filling bedding, clothes, everything, with eggs and filth. Some of the eggs which had been laid earlier and not destroyed had hatched into loathsome maggots.

One day a pleasanter kind of insect was found. The travelers had come upon a hollow cottonwood in which wild bees had secreted a supply of honey. Lee secured some for Nuttall, who felt that if he might possibly swallow and keep down a mixture of honey and water, he could regain some of his lost strength. He had not been able to eat anything but a little hard biscuit and to drink little except tea for days.

When matters seemed to be a trifle improved, they realized with mounting uneasiness that they had come into the country of the Osage Indians. Now and again one was seen skulking in the four-teen-foot horseweed tangles. The trapper and the botanist were alarmed. In an effort to move on so rapidly that they might lose the Indians, they struggled on their way in the heat through tall grasses and giant horseweeds whose leaves and stalks were rough and bristly, but, as they went, they again saw something dodge through the tall growth. Osages were not easy to outwit or to lose.

In the camp that evening, Nuttall again was delirious. A strange coldness and lethargy lay over him a horrid calm which he dimly felt, as if from a far country, must presage death. Unable to lift hand or head, he lay there as if waiting for the end. Yet when morning came he realized with some incredulity that the Indians

had not attacked after all, and he himself was still rather remarkably alive. Somehow, he was hoisted up on his horse, and they
plodded on under an ominously oppressive, cloudy sky until Nuttall, like a rag doll, suddenly slid off to the ground. Again, with a
supreme effort, he was put back in the saddle. The two men continued thus for several interminable days of anguish, in which
Nuttall himself lost track of time. They had expected daily to come
to the Salt River, until Lee finally had to admit that, on top of all
their other troubles, they were lost. It capped their miseries.

They finally came by sheer chance upon the river, and then
worked their way slowly upstream and rested in the lovely, if
sparse, shade of the cottonwoods. A few minutes later, Lee's ailing
horse, which had gone down to drink, sank suddenly into the soft
mud. They could not get it out, could not aid the unfortunate animal as it struggled, whinnying in terror and sinking deeper and
deeper in the viscous mud. As the thick mud and water closed over
the creature's head, it was gone with an ugly sucking sound of
unutterable finality.

Since there was no possibility of obtaining a second horse, Lee
decided that as the only recourse to escape he must make himself
a dugout canoe. He was evidently a resourceful fellow who was
able to do many things. They camped in the shade, therefore, while
the trapper hacked and burned out the length of an old cottonwood log and shaped it for a reasonable degree of navigability.
When Nuttall, unable to drink the nauseous liquid in the river,
crept away from the banks to look for some fresh water, he thought
he saw another Indian, but the vision was gone before he could be
certain of it. Nuttall was feeling better. If the Indians now permitted it, he would surely survive.

Lee in the clumsy canoe, Nuttall riding and trying to parallel
him on the bank, the pair moved on. Now and again as they saw
the smoke from Osage campfires, they momentarily braced themselves to being discovered. Although they moved as quietly as possible, the Indians had no doubt kept track of their whereabouts all
the time; the wonder was that they had waited so long to confront
the white men. Then it finally happened. Nuttall was suddenly surrounded by the Indians. Lee paddled to shore to assist his be-

leaguered friend, who was being ordered rudely by a fat Indian woman to get down from the horse and give it to her. The two men parleyed; they argued; they offered presents. They went when invited—rather, were summoned—to the Osage encampment, where they partook of a mess of cornmeal mush that had been sweetened with pumpkin marmalade. It was a dish which Nuttall could only with the greatest will power manage to swallow and keep down.

Later, when they were about to depart, thinking the Indians had been appeased, they discovered that a dozen Osages had taken charge of the canoe. Lee threatened them. Although he was armed, he was obviously outnumbered and he disliked precipitating trouble by shooting if he could avoid it. The Indians, however, relinquished the canoe, but turned to pilfering the baggage. With great effort Nuttall and Lee at last got away from the thieving Osages, but only by placating them with Nuttall's blanket, several knives, and other pieces of equipment.

But throughout the rest of that nerve-racking day, as Nuttall rode along the weed-grown river bank and Lee paddled the canoe not far from him, the Indians trailed them. The two did not dare to make camp when night fell and a stormy sky presaged rain, but, hoping to elude their followers, continued in the darkness.

Flashes of lightning in the heavy storm which burst upon them showed Nuttall the way across innumerable fords in the winding river, crossings in which he never knew if there might be quicksands or soft mud which would pull him down, horse and all, or knew where the Indians were. Now and again he lost track of Lee's whereabouts, then in relief, by lightning flashes, saw him reappear before darkness and rain hid him again. Nuttall, on one of the crossings, fell off his horse. Gasping, drenched, he struggled to the shore. Although the rain was ending, it was now so chilly in the freshening wind that, whether it was safe or not, he simply had to have a fire. When Lee caught up with him, he reluctantly gathered twigs to build a very small fire, then went off to crouch in the bushes and watch for the Indians. None came.

Shuddering with the chill and the nervous strain of the past hours, Nuttall crouched over the fire. Lee emerged from hiding and cooked some elk meat which had been taken the day before. The storm blew itself out and a late moon arose in a windy sky.

On the next day they decided they must separate. Because Lee and his canoe could go much faster than Nuttall could ride on the bank, the former would continue alone to the Arkansas, while Nuttall, bringing the baggage on the horse, would follow as he could. Lee left him some fire-making equipment and supplies and was on his way before Nuttall found that he could not get a fire started. There was no dry tinder anywhere. The powder in his gun was wet, besides, and it was quite useless either to get some food or to protect him against the Indians. Attacked by mosquitoes, without fire or water and without food, for the meat which Lee had left for him had spoiled in the heat, and he could not have eaten it raw anyway, he spent another dismal night lying among the rank horseweeds on the river bank. It was truly astonishing, he may have let himself ponder in his misery, to what lengths a botanist would go to pursue his career in the search for plants.

When he felt he could not endure another day of the grueling torment of plodding travel, he suddenly saw ahead of him, looking as splendid as a palace in the golden light of the setting sun, the log walls and stockade of Mr. Bougie's trading post. It was an "asylum which, probably, at this time, rescued me from death. My feet and legs were so swelled, in consequence of weakness and exposure to extreme heat and cold, that it was necessary to cut off my pantaloons, and at night both my hands and feet were affected in the most violent cramp. . . . I remained about a week with Mr. Bougie, in a very feeble state, again visited by fever, and a kind of horrific delirium, which perpetually dwelt upon the scene of past sufferings."

Sufferings or not, Nuttall still had his precious plant specimens. They were the first to have been collected in the upper country of the Arkansas River.

Not until October was he able to travel and make his way to the Mississippi, go by boat to New Orleans, and by ship around to Boston where his botanical friends, in their comfortable city homes and not realizing his abject hardships in the name of science, were impatiently awaiting his new collections and his list of new species.

It was time that he did so. He must settle down to his position at Harvard.

Although Thomas Nuttall could do this for a time, the West forever called to him. His book of ornithology was published in 1834, but in order to complete his work on American plants, he had to get himself all the way to the Pacific Ocean. Although he had made two journeys in that direction, he had not as yet seen the Rocky Mountains or the West Coast. He could wait awhile, but he would have to go there someday.

As soon as his bird book was published, therefore, he and John Kirk Townsend in 1834 set off with the Wyeth expedition for the

Nuttall's Pacific Dogwood

Columbia River,* and Thomas Nuttall at last had his wish to collect the trees and wild flowers and ferns of that alluring country. To top it all, he went on to Hawaii before finally returning home, by way of Cape Horn to Boston, to complete his book of flowers and trees of North America. By the time most of the other westward-traveling naturalists had reached the Pacific, they would realize that the indefatigable Nuttall had collected there long before them.

* See Chapter 13, *Men, Birds and Adventure,* by Virginia S. Eifert.

8. Mystery of the Mississippi

It was a very long river. It bisected a continent and influenced a nation, but, by 1820, no one had as yet learned how long it really was, nor where it had its origin. Knowledge of the Mississippi had unfolded piecemeal. Every explorer, each in his own way, had seen part of the river and part of its wildlife and reported it as he saw it, according to his own observation and in the enlightenment of the times. The situation was something like that of the Blind Men of Hindostan who went to see the elephant. Each comprehended the wonder in a different way, but each of them, according to his viewpoint, had indeed seen the elephant.

Every man who touched on the river at some point in its several thousand miles of winding, southbound course, had obtained a very different idea about what this stream was and about what lived in and beside it. This was a river of willows, a river of high cliffs, a river of broad valleys, a river of majesty, a river of monotony, a river of drama, above all a river of mystery.

For a long time after De Soto in 1541 came to the banks of the Mississippi and saw its waters, and long after the river itself had conquered the conqueror, little was known about the vast water-

way. Spaniards, acquainted only with the area around the mouth
and little more than two hundred miles upstream, came and went.
In 1673, two canoes paddled by Frenchmen came out of Canada.
They passed through the Great Lakes and navigated down the
Wisconsin River to enter the upper Mississippi. They had heard
stories from the Indians about a certain great river flowing southward below the lakes; there was now a growing urgency in France
to know quickly where that river actually emptied and just what
and who occupied its shores. Was this, they wondered with some
worriment, the same River Espiritu Santo of the Spanish, or was it
another one, possibly that long-sought way to the South Sea,
emptying into the Pacific Ocean instead of the Gulf of Mexico?
When Louis Jolliet was commissioned by Louis XIV to go and find
out and to map the way as well as take notes on the natural history
and Indians along the route, white men finally began to be aware
of what the Mississippi might hold in store for them.

Jolliet took with him five Frenchmen, and at Michilimackinac
had picked up the eager Father Jacques Marquette, who longed
to carry the word of God to the western places. Both Jolliet and
Marquette were naturalists. Even though they frequently did not
know the identity of what they were seeing, they nevertheless
were good observers. Marquette was particularly interested in
botany. From the two we have what was no doubt the earliest
account of the wildlife of the Mississippi.

It was a fine day in June when they reached the willow-grown
sandbars where the Wisconsin River flowed into the Mississippi.
To Marquette and Jolliet it was an exciting and an almost holy experience as they floated at last down the beautiful Mississippi with
the pristine landscape of the hills and, filling the bottom-land,
marshes full of birds and plants, and devoid, it would seem, of all
danger or menace. Marquette wrote:

There are woods on the two sides as far as the sea. The mightiest of
the trees that one sees here are species of cotton wood, which are unusually stout and lofty; wherefore the savages use them to construct
canoes, all of one piece, 50 feet long and 3 feet wide, in which 30 men
with their baggage can embark. . . . That soil is so fertile that they raise
corn three times a year. It produces spontaneously fruits which are un-

known to us, but which are excellent. . . . There are hollies and trees of
which the bark is white . . . papaws, which is a small fruit that is not at
all known in Europe . . . turkeys everywhere, parrots in flocks, and
quail. . . .

Marquette and Jolliet went as far as the mouth of the Arkansas.
After some unpleasant experiences with the Quapaws, and hearing
from them that the Spanish were at the mouth of the Mississippi,
several days' distance, they turned about and paddled laboriously
upstream. The battered birch-bark canoes returned by way of the
Illinois River to the lakes. Jolliet and Marquette had discovered
that the great River Mesesebe was actually connected with and
part of that same river which the Spanish knew; had found that it
did not after all empty into the Pacific as the king had so fondly
hoped. With reports of wildlife and Indians, the expedition made
its way back to Québec.

A few years later, while the Sieur de La Salle took an expedition
to the mouth of the Mississippi and claimed it all for France, one
of his band, a Recollect priest called Father Hennepin, on a minor
expedition of his own, worked northward up the Mississippi. He
traveled from the mouth of the Illinois, made a good many inept
mistakes, was captured by the Sioux, and was taken by them as far
as the Falls of St. Anthony (St. Paul and Minneapolis) before his
captors hauled him inland to Lac Milles Lacs in Minnesota.

For some time after that, hardly anything further was known
about that upper river. Although Louis Jolliet had surmised that
it had its origin in various lakes of the north, no one really knew
where it began. In 1700, Du Charleville at Mobile, hoping to ex-
tend French trade to the unknown source, went north by canoe in
an effort to find it. However, having come more than eighteen
hundred miles to the Falls of St. Anthony, and becoming greatly
disheartened by the appalling endlessness of this formidable
stream, he was further discouraged by the Sioux, who did not want
him trespassing in their country. They informed him that the Mississippi extended to the north for as great a distance as it had
already come. Gloomily deciding that the river must surely orig-
inate in the Frozen Ocean, he turned back in defeat.

A little later, the young Frenchman named Simon Le Page Du

Pratz, coming to Louisiana on a land scheme which was part of the infamous Mississippi Bubble, and being something of a naturalist, compiled a book about his discoveries along the Mississippi as far north as the Ohio. Although his natural history drawings were crude when they are compared with the beautifully detailed paintings of Mark Catesby who, nearly at the same time, was collecting and painting wildlife in Carolina and Virginia, Du Pratz tried to delineate what he was seeing. He described a great many plants, birds, and other creatures of the lower river, but, after he had gone back to France, hardly anyone else troubled himself about these mysteries of the Mississippi. *

No one, really, knew very much about anything that dwelt in the upper reaches of the river. Then, when the Minnesota country, following the Louisiana Purchase, became annexed to the young United States in 1803, it became both more possible as well as more imperative to discover what lay northward, and to determine the exact source of the Mississippi. For in 1803 there were three great questions asked by Americans, questions which must be answered as soon as possible: where the Missouri River had its beginning; how close it came to the headwaters of the Columbia; and the exact location of the source of the Mississippi. The latter, in determining whether it was within the territory of the United States or lay in Canada, might be vitally important to future peace and the handling of international relations.

Shortly after President Jefferson had sent Lewis and Clark up the Missouri River to find the answers to the first two questions, he dispatched young Lieutenant Zebulon Pike from St. Louis to take a party of soldiers up the Mississippi and find the headwaters. Although Lewis and Clark were perceptive men who were both interested in natural history and had, in fact, been ordered by the President and various men of science in the East to take particular note of the wildlife, Pike was neither a naturalist nor was he interested in the subject. On his expeditions, he seemed to have all he could do to keep his men alive and fed, get them through a difficult country, and deal with the Indians and other problems. He

---

* See Chapter 4, *Men, Birds and Adventure,* by Virginia S. Eifert.

had no time to be concerned with plants and animals, except those which might be eaten.

Zebulon Pike, who departed far too late in the season, reached the upper Mississippi in midwinter. On Christmas Eve, at the location of Brainerd, Minnesota, he was in a bleak and terrible wilderness of great cold and wind. Finally attaining the American Fur Company's post at Sandy Lake, two miles from the narrowing river, Pike left most of his men to recuperate and, with several Chippewa guides and a few soldiers, pushed on. He was obsessed now with reaching the end of this infernally long and preposterously winding river.

In the upper country, the Mississippi spreads into lakes and then it narrows again. There are confusing forks and many tributaries which feed its growing power. Pike, unaware of the true nature of the Mississippi, took the wrong fork in following the icebound course. Ultimately he found himself on the frozen shores of Leech Lake. The river, he believed, must have its origin in this lake— this, surely, must be the headwaters of the great Mississippi. The Indians who were with him, impatient to return to where it was warm at the fur post, hastily assured Pike that this was indeed the source he sought, and Pike was eager enough and sufficiently convinced to believe it. On February 1, 1806, he planted the American flag on the shore of Leech Lake.

"I will not attempt to describe my feeling on the accomplishment of my voyage, for this is the main source of the Mississippi," he wrote happily in his journal.

But it was not.

He may have believed his own statement, but some men were not convinced. One of these, some years later, was young Henry Rowe Schoolcraft, who had been thirteen years old when Pike announced his discovery. Schoolcraft, who became a geologist in New York, had visited the lead mines on the Mississippi near Dubuque and Sainte Genevieve, and he now thought he knew something of the river. Disbelieving the authenticity of Pike's conclusion, he longed to go and seek out the real source.

Soon after Michigan became a territory and had a governor, the very stout and jovial politician and veteran of the War of 1812,

General Lewis Cass, it became necessary to know something of how much land the territory encompassed. It was needful to know about the Indians living there, as well as to amass information on plants and animals, on topography and on mining possibilities. The Indians and trappers had spoken of copper mines along Lake Superior; there were also rumors of silver deposits. Now that it was a United States territory and would eventually become a state, or several states, it was certainly time to know more about that whole area.

The Cass expedition went well equipped with men, brains, and supplies. The Chippewas of Lake Huron, some of the best canoe-

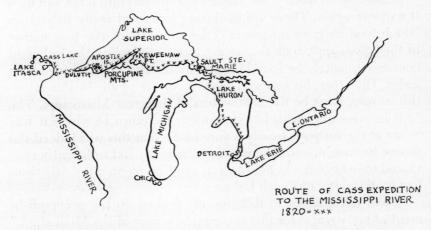

ROUTE OF CASS EXPEDITION
TO THE MISSISSIPPI RIVER
1820 = ×××

makers in America, had built three superb birch canoes to carry the entire expedition. Thirty feet long and six feet wide, these monster craft would travel easily with about four tons in each when loaded with men and baggage. Eight men, French-Canadians who had been hired to paddle each canoe, would move swiftly and rhythmically, averaging four miles an hour. This was the grueling pace of the men and canoes on fur brigades in the North. They were accustomed to it, but it was often hard on Cass and the officers.

The Cass expedition included Henry Rowe Schoolcraft as geologist. He was at last, after years of dreaming, going out to the Mississippi. On May 24, 1820, the canoes set off, the whiskery voyageurs at the paddles, and with Indian guides, soldiers from the fort at Detroit, and a botanist, Captain David Bates Douglass (not the David Douglas who appears in the next chapter). There were several topographers, a meteorologist, and several army officers. Last to be added to the Chippewa canoe was the great bulk of Lewis Cass, the governor himself, whose weight made the craft go down alarmingly until some of the baggage was shifted.

Considering the stops which had to be made when the lakes were too stormy for the canoes, or when the men had to halt to talk to Indians and confer with the leaders, it was a rapid trip. There was all too little time in which to make a detailed exploration. Any natural history collections had to be made quickly at the shore stops and the night camps. The expedition thus moved through Lake Huron, paused for supplies at the fort on Mackinac Island, then went up into the St. Mary's River at Sault Ste. Marie and on to Johnston's trading post and fur station beside the Sault.

Here they rested while the Indians and boatmen expertly portaged the great canoes and all the baggage around the falls and reloaded them beside the red rocks where the river flowed out of Lake Superior. Johnston, a knowledgeable man, talked of the Indians, of animals and plants, and of the lakes, and introduced to the officers his charming daughter Jane. Her mother was a Chippewa, and Jane herself was a well-educated and delightful person. Schoolcraft became very much attracted to her and rather disliked having to leave so soon; he promised to come back. Jane Johnston at the Sault was not someone he could forget easily. Her father sent along several of his best Chippewas as guides to be mediators with some of the more recalcitrant Indians ahead. The guides smoothed the way more than once for Cass's party in often delicate situations which, without them, might have led to disastrous consequences for the entire expedition.

They were launched at last upon Lake Superior—Lake Superior with its pale-blue, chilly waters and its strange mirages which formed every afternoon above the far blue horizon. The travelers passed the outthrust of the mountainous Keweenaw Peninsula and made several stops there to investigate rumored copper deposits. They came along the rocky southern shore above which, tall and dark, rose the bulk of the Porcupine Mountains. The lure of the mountains and their unknown heights drew the minds of the men to contemplate what they were missing by going on and not exploring in detail. The whole area needed to be examined, but there was not the time just now for this; not now. There was, however, time for a short side excursion up the Ontonagon River.

Beaching the canoes a little distance up the rocky river, the men set out on foot for an exploration of its course, for there had been rumors from Indians and trappers about a great mass of pure copper lying in a river, and Cass and the others, thinking it might indeed be in the Ontonagon, were impelled to investigate. This was also an opportunity for the botanist to collect plants and for Schoolcraft to study the geology of the immediate area.

It had been painfully tantalizing more than once, both for Douglass and for Schoolcraft, to pass along these shores and only skirt that splendid dark wilderness. They, however, on the brief shore stops, had already managed to collect a good deal. They had found the round-leaved pyrola blossoming in pine woods on the Keweenaw Peninsula, had seen *Potentilla tridentata* with its glossy leaves and white flowers growing compactly in crevices of rock, and an attractive, evergreen-leaved, creeping plant called *kinnikinick* by the Indians, which, the Chippewa guides assured them, was used for tobacco. Since, in the Algonquin language, the word *kinnikinick* designated any mixture of leaves which could be smoked, the name was neither a reliable nor a usable one.

The identity of the glossy-leaved plant puzzled young Douglass. He could not find it in Pursh's book of American botany, nor in any description to be found in Michaux and Eaton, the two great authorities on American plants. Yet this creeping plant must have been the bearberry, *Arctostaphylos Uva-ursi,* which had been named by Linnaeus himself. It was common to the rocky places, to the sandy places, to the heights of mountains and to other punished habitats. If the men of the Cass expedition had climbed up through the dark-shadowed old hemlocks to the crest of the Porcupines, they would have found it growing abundantly on the conglomerates overlooking Lake of the Clouds, as it grows today.

The enthusiasm of Schoolcraft and Douglass must have been infectious. After the former had roamed along the rocky shores and pebbly beaches of Lake Superior and had picked up carnelians and bits of copper, greenstones, fossils, and other interesting specimens, some of the voyageurs themselves began to look and collect. For years they had known this lake, but they had never seen anything very much along it until now. They set themselves to look for plants for Douglass's collection. Even when they came with something well known and already represented in his material, he appreciated the men's earnest efforts and encouraged them. There

Bearberry

was always the chance that they might one day bring him something new and exciting and momentous, to make it all worthwhile.

Governor Cass, however, many times may have regretted his coming here. He suffered in the heat and hardship of the long daily grinds and found little comfort on the rocky shores at night. Considering his bulk and his dignity, however, he maintained himself very well. The morale of the whole expedition was high until that day when they went up the Ontonagon River to search for the copper mines and for the great nugget.

The temperature stood at ninety degrees, and the weather was very humid. Even in the dark shade of the hemlock and maple forests covering all of that southern shore, the heat was almost sickeningly oppressive. Governor Cass, streaming with perspiration and his face alarmingly scarlet, puffed and groaned and stumbled and gasped, but he kept up fairly well with the rest.

In the bottom of the precipitous gorge of the river, whose tall, earthy and rocky banks were continually crumbling and falling down, they came upon a massive piece of tarnished and corroded copper which had been reported so long before by the explorer Alexander Henry and by later trappers and Indians. It was an immense thing, believed half a ton in weight, lying half submerged in the river, but, because of its weight, quite impossible to bring out whole. The landscape was so desolate, broken, and inaccessible that such an attempt was unthinkable. Schoolcraft, describing the place, wrote:

These bluffs may be considered one hundred and fifty feet in perpendicular height, and are capped by a forest of pine, hemlock, cedar, and oak. On the right hand, partly immersed in water, reposes the copper rock; on the left the little island of cedars divides the river into two channels. . . . The masses of fallen earth,—the blasted trees, which either lie prostrate at the foot of the bluffs, or hang in a threatening posture above,—the elevation of the banks,—the rapidity and noise of the stream, present such a mixed character of wildness, ruin, and sterility, as to render it one of the most rugged views in nature. One cannot help fancying that he has gone to the ends of the earth, and beyond the boundaries appointed for the residence of man. . . .

The party then scattered for wider coverage of the area. While the naturalists explored in one direction, the governor and some of the others went elsewhere; they were all to meet again before dark at a certain spot on the river. The first group had returned well ahead of time to the rendezvous, for the heat had been too great to make climbing especially pleasurable or even endurable. They made a small fire to cook their supper, and waited. The governor and his party did not come. The others conjectured a variety of reasons for the delay, then grew more worried as night darkened, and they began to fear that Cass had fallen on the rocks, or had suffered an apoplectic stroke, or had died of the heat. Schoolcraft and Douglass searched in the dark forest on the bluffs; they called, shouted, but heard nothing except the hollow sound of running water and the squeaks of the bats. A deer snorted suddenly and thundered off, startling the searchers. The forest was black and mysterious and now was still, all but the thin whirring of innumerable mosquitoes and a screech owl which made a purring sound somewhere in the hemlocks. A wolf howled, and they heard a running sound and a desperate cry from an animal being caught and killed, and then it was all very quiet again.

The men fired off their guns, but only echoes replied. To lose the Governor of Michigan Territory—it was an awful thought. Their courage low, the men felt very much alone around the fire with that black night enclosing them beside the Stygian Ontonagon and its gorges. They waited. Captain Douglass silently arranged specimens in his plant press. Schoolcraft didn't have the enthusiasm even to look at his minerals. The men slapped mosquitoes and waited.

Finally, one said he would take a canoe farther up the river to see if the governor might not be there. It was an excellent piece of intuition, for, as he rounded a bend, he saw a small fire up ahead and there found Lewis Cass and the others sitting disconsolately and quite worn out and hungry. They had been lost, Cass feebly explained, with some difficulty tottering to his feet, only about three miles from the river. The forest had been so dense that even the Indian guide himself had soon become quite lost. Almost exhausted, the governor and his companions clambered into the canoe and were conveyed back to the camp.

Next day, revived by rest and food and rejuvenated by the cooler air of the morning, they were on their way again in Lake Superior, the forests of the Porcupines on their left, the great expanse of pale lake water on the right. Following the curving shore, they went past the amber-clear Presque Isle River tumbling down from its series of falls and cascades. Circling the curve of Chequamegon Bay, they paused on Madeline Island in the Apostles for rest and supplies at Michel Cadotte's fur establishment before rounding the Siskiwit Peninsula. The canoes headed northwestward on the long haul to the Fond du Lac where Duluth and Superior now stand. Leaving the canoes here, they struggled on foot along the rocky course of the St. Louis River westward, finally working their painful way through what seemed to be endless sphagnum bogs and spruce swamps. The botanist paused many times to collect small plants growing in the bog moss, but, to the rest of the men, their exhaustion growing in the spongy, enervating footing and the steamy heat, it was a tribulation. They finally had to aid the governor over the last portion. It was a huge relief to reach the haven of the American Fur Company's post at Sandy Lake. It was an even greater relief, after all the way they had come, to know that they were now only two miles from the Mississippi River.

Schoolcraft was growing more and more eager to get on the Mississippi and to pursue it to its ultimate source, but Governor Cass now was inclined to take his time and to delay further by having endless conferences with the Indians. He had had almost all he could take of exploring, had reached a point of not greatly caring whether or not he ever saw the source of the Mississippi. Where a river started wasn't so important as where it went, or what that stream amounted to farther on. The Mississippi had certainly left nothing to be complained of in its downstream accomplishments. Let Pike's source continue to be Leech Lake. What, after all, did it matter? Who cared? The source was undeniably in American territory and that was really all that the government wanted to know.

Schoolcraft, however, was one who wanted to know the truth. So did some of the others. They could not bear to go back now. The geologist argued, and finally Cass wearily agreed to continue. On the river with smaller canoes than those left on Lake Superior, they now moved strongly against the current until they had reached the forks. The left-hand fork was the one which Pike had followed. The expedition, therefore, keeping to the right, came into the broad, shallow waters of Lake Winnibigoshish. Seeking an outlet, they paddled along the rice-grown rim and found several streams running into it. The largest of these was chosen as possibly being the Mississippi; no one could be certain. They paddled on a few more miles until the canoes, pushing through the wild rice beds, floated into still another lake.

It was a beautiful, clear body of water about eight miles long; it was gravelly and had sandy shores; pine forests circled it. In this place there was a good feeling of finality. It was a suitable source for any river. Cass was so convinced that this was indeed the Mississippi's origin that, even after some of the men had paddled around the circumference of the lake and reported two small streams entering at the far end, he still would not change his mind.

As far as Lewis Cass was concerned, this was as far as he was go-ing, so this was therefore obliged to be the source of the Mississippi River. He was gracious in letting Douglass, Schoolcraft and the officers name the lake appropriately in his honor. They called it Cassina Lake, with an official flag-raising ceremony. Today it is known as Cass Lake.

But it was not the source of the Mississippi.

The mission was accomplished. The expedition started back. Yet, Henry Schoolcraft and some others could not feel completely satisfied. They knew that on the far side of Cassina Lake there had been a river . . . knew that the Indians had said that it flowed from a body of water the trappers named Lac La Beesh some forty miles away. But there had been no time to investigate. They were down-bound now, following the river for a distance, seeing how willows grew along both shores as they grew all the way down to the Gulf of Mexico—the Mississippi was indeed a river of willows extending all the way to the sea and filling the valley with their greenery and their own special fragrance.

Downstream, the party disbanded. Schoolcraft went to the lead mines at Dubuque, then traveled overland by way of Chicago to Detroit. September 13, 1820, and it was over . . . finished. The Cass expedition, the world had been told, had truly discovered the head-waters of the Mississippi. But some of the men, especially School-craft, knew that they should have gone on.

It had, however, been highly successful as the first scientific ex-pedition into the upper Great Lakes and the upper Mississippi. Numerous mineral and rock specimens had been brought back, many Indian artifacts and abundance of fossils, as well as the first collection of living shells of the cold blue northwest lakes and the mysterious upper river. And although Captain Douglass had not been able to devote as much time as he might have to his plant collecting, he had really done very well. His whole array of pressed plants and seeds had been sent at once to New York to be exam-ined and classified by Dr. John Torrey who found, with some grati-fication, that Douglass, besides having been the first botanist to collect in the upper Great Lakes, had in fact discovered some plants which were new to science.

Well, it was finished. Schoolcraft would have to be satisfied with what had been accomplished. But when, in 1822, he was appointed by Cass as Indian agent in charge of tribes of the Lake Superior region, with headquarters at Sault Sainte Marie, he was considerably closer to that part of the country, which continued to draw both his attention and his thought, as well as his affections. The charming Miss Jane Johnston was waiting for him. Improving the acquaintance, he married her in the following year.

The lovely and intelligent Jane, linked both with the world of white culture and to the world of her mother's people and their own culture, with her knowledge of Indian customs and language, and her contacts with her mother's relatives, was a tremendous help to her husband in his work with the Indians. She was his assistant in the later project of writing a great series of books preserving the ethnology of the Indians of the Great Lakes.

Engrossed in work as he was, the years went on and he somehow never had the opportunity to go back to explore the rest of the tantalizing Mississippi River.

Top of the Mississippi

In 1832, needing to go westward in an attempt to settle some trouble between the Chippewa and the Sioux, who were forever at odds with each other, Schoolcraft decided at the same time to visit the trading posts of the area and note what the Hudson Bay people were doing both for the Indians and against them. He took with him an engineer to make a detailed map; a missionary, the Reverend William Boutwell to carry the gospel to the Indians; and a physician-naturalist, Dr. Douglas Houghton, who was not only going along to vaccinate the Indians, but to collect plants and other natural history specimens. In 1832, Houghton was only a young man of twenty-three. He welcomed the chance to get away from the drudgery of private practice, preferring, he said, "the pleasure instead of a mental feast upon the hidden treasures of nature."

Lieutenant James Allen came along as map maker and journal keeper, and there were soldiers for protection. George Johnston, Jane's brother, came as interpreter and brought with him a capable Chippewa guide, Osa Windib, called the Yellowhead, who had perhaps been named for the yellow-headed blackbirds flashing in the marshes. Osa Windib assured Schoolcraft that he knew the way to the little lake called La Beesh. For, although the expedition was ostensibly for other purposes, Schoolcraft was going to use it to find the source. He had to know.

Immensely eager to get started, and well outfitted, they were on their way through the Sault, into Lake Superior again, heading inland from the Fond du Lac, going overland through the bogs, once more reaching Sandy Lake, and traversing the two miles to the Mississippi. The journey was, of course, not quite so rapid a transit as this account makes it sound; continually, all along the way, they were obliged to meet and confer with the Indians, to placate and vaccinate, to collect specimens of wildlife and to record in notebooks all that they found and saw and did, and to map the way besides.

Reaching Cassina Lake, the Yellowhead led the way around the sandy, curving, pine-dark shores to a tributary entering the western end; this, said the Yellowhead, was the Mississippi. They paddled the unknown stream and found it more and more winding. The river, to the astonishment of Schoolcraft, Allen, and Houghton, was forming an enormous fishhook; it was no longer coming out of the north, as everyone had previously assumed, but was curving strongly from the west around to the south. This was incredible. It was as if the river, having gone as far north as it cared to go, was now trying to tie a knot in itself.

After passing through Bemidji Lake, they left the canoes on the shore. The Indian said that the river was now too small and too shallow at this time of year to accommodate them. He led the way through the willows and tamaracks in a marshy bottomland, pushed through tangles of alders and into a wild rice slough. The men came out suddenly into the sunshine of a lovely little lake on which a black and white loon was swimming, and around the shores the blackbirds swung on cattails and sang in the sunshine. The waters shimmered in the July brilliance. A muskrat swam away from the place where a small, clear stream came tumbling languidly over gravel and stones, heading out of the lake and on its way to the sea. And so this was it . . . this was the birthplace of the Mississippi.

There was, somehow, no questioning it. At 1,475 feet in altitude, the river at their feet was pouring on its way down 2,552 miles to meet sea level at the Gulf of Mexico. Nearly three hundred years after De Soto had first seen the river, the entire course of the Mississippi had finally made itself known to men.

There needed to be an appropriate name for this important little lake. Several days before, en route, Schoolcraft had discussed it with the missionary. He wanted a name which was unique. It must not be that of a person who, a century hence, might be forgotten, but a title of its own, a title with stature, one which was as special and as full of meaning as this lake itself. He questioned what the Latin might be for *true source*. Boutwell thought for a bit, and then decided that *veritas caput,* or true head, might be close enough to the meaning he sought.

This was too cumbersome. Schoolcraft then contrived a lopping off of letters. He omitted the first three and the latter three, and found that what remained had become Itasca. Itasca Lake—this is the name of the body of water which is acknowledged as the source of the Mississippi.

Dr. Houghton botanized around the headwaters. He collected ferns and wintergreen and fly honeysuckle and Canada anemones in the red pine forest, found dwarf birch, tamarack, swamp spruce, wild calla, wild blue flag, and Jo Pye weed around the small river itself. He catalogued the willows—the black, withe, yellow, almond-leaved—and wondered how many of them followed along the entire length of the Mississippi. He also found a new dewberry which later was named for Schoolcraft, and Dr. Torrey chose to name a new goldenrod and a new sedge in honor of their discoverer, Douglas Houghton.

Then they were on their way once more. After all, it had not been the sole purpose of this trip merely to discover the origin of the biggest river in North America, and to locate a few new flowers. But of all the other things which took place on that journey in 1832, it was the discovery and the naming of Lake Itasca, several new plants, and the solution to the mystery of the Mississippi which most people remember today.

Houghton's Sedge

Douglas Fir

9. David Douglas

For six long and tedious weeks, the *William and Ann* had waited outside the pounding waves and cantankerous currents at the mouth of the Columbia River before the vessel dared to move into the river itself. Since leaving a latitude a few degrees north of the equator, the ship had been beaten and tossed on storms in the Pacific. David Douglas was sick to death of seas and storms. He was hungry for greenery, hungry for trees, most of all was half starved for a sight of some new plants. Having come all the way from Scotland to explore the Columbia River country, David Douglas had resented the wasted time as the ship had fought and waited, had approached and retreated, then fought again to pass the great sand bar lying in the bay which formed the entrance to the river.

As the *William and Ann,* seeking the best anchorage, finally moved slowly into quiet waters, Douglas and his friend Dr. John Scouler leaned on the rail and let their eyes follow along the shore to discover what was to be seen even before they had landed. It was 1824 and they were the first naturalists to visit the Columbia since Dr. Archibald Menzies, on Captain George Vancouver's expedition in 1792, had had so little opportunity to explore and collect in the wonderful Northwest Coast country of America.

The ship drifted past a massive and splendid wall of trees. Lush, dark green, glorious in primitive grandeur, they rose from banks of ferns and masses of shrubbery covered with dark green and glossy leaves. Now and again a bird sang, but the bird voices, like the trees, the ferns, and the bushes, were foreign to David Douglas. Then he thought he recognized hemlocks; he had known them in New England several years ago, admired their dark and drooping boughs which made them one of the most graceful trees in the world. He thought he also recognized the tall pyramids of tapering balsam firs, even believed he detected their splendid and invigorating perfume. But with so much that was fragrant and exciting to smell, after only sea odors for so long, he could not be sure.

Aside from the two species he thought he recognized, there was another kind of conifer which grew in the wall of trees, a tree which certainly was the most majestic plant he had seen anywhere on his travels. It was truly a king of conifers, very tall—immensely tall—its great bole rising in a gradual diminishment from a massive lower trunk to a thin tip standing several hundred feet above the sword ferns. Extending out in a curve, the boughs had long, drooping twigs covered with thick, gray-green, short needles. The younger trees along the river were branched all the way to the ground, but where the shade below was very dark, the lower branches of the greatest of the trees had been naturally pruned, leaving an enormous, smooth trunk whose boughs were only very high in the crown.

He was still wondering about them as the ship anchored in Baker's Bay. As Douglas and Scouler in a small boat hastened ashore, they were delighted to come at once upon some of the very plants which Dr. Menzies had collected—the handsome salmonberry with its white flowers which looked like single wild roses, and thickets of salal, or shallon, with their thick, oval, evergreen leaves. Douglas was happy. He couldn't remember when he had been more thoroughly and warmly fulfilled and happy. Here not only were Menzies' discoveries but, added to them, were these monumental conifers and undoubted numbers of other new and exciting plants.

The great trees fascinated him. Menzies had indeed described them, but something had happened to his specimens and so the tree had never been given a name. Although David Douglas did not know it then, this would be the tree which would bear his own name—this was the Douglas fir, *Pseudotsuga taxifolia*, seen on three expeditions along the shores of Oregon before it finally was given a name.

It had taken him eight and one-half months of a long and perilous sea voyage, as well as all of his twenty-six years, to reach this point and to stand here at last on the soft, resilient, red-brown earth of a primeval forest. The fact that he was here at all beneath these astonishing trees along the Columbia River was the end result of factors which had been set in motion long before he was born. David Douglas's presence beside the Columbia was the climax of a chain reaction which had passed through at least two previous expeditions, the first one of which had taken place fifty years before.

Perhaps it had all begun in 1769 with the astronomical phenomenon of the passage of the planet Venus across the sun. Captain James Cook had been sent out by the Royal Society of London to observe this important transit. At the same time he was to explore and map the South Pacific islands and to search for the location of the rumored and mysterious Southern Continent. As the first purely scientific expedition, it was planned and carried out with care.

On this momentous journey had gone the young Sir Joseph Banks, a very wealthy amateur naturalist who brought with him an entourage which included an excellent Swedish botanist, Dr. Daniel Solander, who had worked with Linnaeus; two artists; a secretary; four servants, and his favorite greyhound. Banks's one reason for leaving the easy life of a wealthy gentleman was his urgent inner drive to see, discover, and learn about the wildlife of the world. To go on Cook's expedition in the Southern Hemisphere was an opportunity not to be missed. To his friends and relatives who protested this rash plan, he assured them that he would rather die happily exploring in the South Seas and finding new plants and birds than to live long and fatly, as an English gentleman of means was expected to do. Young Sir Joseph, departing joyfully for the South Pacific, was gone from England for nearly four years.

The Cook expedition explored portions of the unknown lands of New Zealand and Australia which had never before been penetrated, and found some of the most curious plants and animals in the world. On the Australian coast the party named a place Botany Bay because of the wealth of new plant specimens, including several species of eucalyptus, which Banks and Solander found there. Not far from here they discovered a most remarkable and astonishing beast which went off in exaggerated leaps like some nightmarish rabbit of gargantuan proportions. Its speed was such that the men decided to pit one of the strange animals in a race with the greyhound . . . but the kangaroo won the race. Even a fleet greyhound could not compete with one of these masterful runners and leapers of the Australian bush.

Vancouveria

Everywhere, on island after island, there were the exciting new plants to be collected—live plants, seeds, pressed specimens—and often in a sort of botanical exchange Banks and Solander left English garden seeds to be planted in the earth of those far Pacific islands.

Eighteen years later, when Captain George Vancouver was sent on an exploration to the Northwest Coast of America, via some of the same Pacific islands known by Cook and Banks, Sir Joseph had a request to make. He had known Vancouver long before and had not liked him particularly as a person; nevertheless, he was respected as an excellent seaman who had learned his skills from the incomparable Cook himself. Joseph Banks could not endure the thought of an expedition's going out anywhere in the fascinating world without a naturalist aboard. It was a waste of equipment, men, and exploration unless as much as possible were learned about every land touched upon. Therefore, the Royal Society, urged by Sir Joseph Banks, appointed Dr. Archibald Menzies to go

Menziesia

with Vancouver and collect plants and seeds, along with specimens of every other sort of wildlife he came upon and could make off with.

Into the unexplored Northwest Coast wilderness of North America, thereupon, Archibald Menzies had walked as the first naturalist to find and collect some of the incredible plants of that area. He and Vancouver went up the coast as far as Cook Inlet in Alaska, seeking the fabled Northwest Passage, then came down to California. Menzies brought aboard ship a splendid array of new trees and flowers, many of them in tubs and containers to be transported home alive. He had not, however, been able to spend time in going very far up the Columbia or very far inland, particularly after he and Vancouver had quarreled bitterly over the incessant specimen-getting and the care it took to maintain the living plants on shipboard. There had been so many places which Menzies wished to see and to explore, so many plants which he had failed to bring back. It was a pity to have missed such glowing opportunities, but Archibald Menzies, under the circumstances, had done the best that he could.

Thirty years after Menzies visited the Northwest Pacific Coast, Sir William Jackson Hooker sent David Douglas to the same area, and for the same purpose. Hooker was a friend of Banks's and a member of the Royal Society; Hooker's special protégé was the young Douglas, who vastly admired Banks, Menzies, and Hooker.

As a culmination to and inheritor of adventure left undone by those other botanists, he was here at last, David Douglas, on an April day in 1825. A very small human being standing dwarfed in the midst of huge and silent trees, yet he did not feel out of place or fearful. Trees were his life. Deeply and happily, they gave him a sense of goodness and fulfillment. In the dramatic shafts of sunlight striking through the forest he was eons away from the smoky atmosphere he had left behind in Glasgow and London less than nine months before.

David Douglas had been born in Scone, the ancient seat of Scotland's rulers, on June 25, 1799. A poor student in school, he had constant conflicts with his teachers, and was happy only when he was outdoors looking for the plants, the birds, and the small animals of the hills and moors. Because he had had to walk six miles to school—six lovely, exciting miles along which new things were to be found every inch of the way—it was no wonder that he was so often late to classes, or else forgot to go entirely. When school discipline was administered in the form of a blistering chastisement with a leather strap split into narrow strips so that it would hurt over a wider area, young David simply set his round face stolidly. With his pale red hair standing up, blue eyes blank, he endured it. When it was over, he went back to his birds and flowers.

Although he was the despair of his teachers and his parents, he shone in the subjects which had meaning for him, the subjects which had a bearing on natural history. In this he read everything he could find, together with stories of adventure and exploration; he longed desperately to be an explorer. He would then see all the trees and flowers in the world, would see all of them and learn each one's name.

Since he was obviously never going to have either the educational background or the money to take him to college, young David left school at about the age of eleven and went to work. It was heavenly work—his father had used his friendship with the head gardener at Scone Palace to get the boy a job as apprentice. Thus the gardens were the true beginning of David Douglas's education. This was the kind of education which would be part of him and, unlike the lessons avoided at school, would be soaked up with an endless eagerness and thirst. He also studied botany from the gardener under whom he worked, drained his mentor of all he knew, and then eagerly sought more. He learned the names, studied plant processes, went to the head gardener himself for books and information. After hours and on holidays, he and some of his new friends went on botanical-collection excursions nearby. The trips made him yearn all the more for a real expedition to remote places in the world.

When he was eighteen, he obtained a position in the gardens of Sir Robert Preston near Dunfermline, where he had access to a superb botanical library which Sir Robert kindly opened for his use. For to the youth botany had become a challenge, and it was not limited to the contents of a garden. Botany encompassed the whole world, and he was determined to make that world his own.

The next step in this direction was a position at Glasgow in what certainly was not an ordinary location, for this was in none other than the famous Botanical Gardens. There, inevitably, as if this had been intended from the beginning, David Douglas met the eminent Dr. William Jackson Hooker, Professor of Botany at Glasgow University. When he came to the Botanical Gardens to give evening lectures for his medical students and others who came from far places to listen to him, David attended the lectures.

Hooker's personal magnetism filled his lecture room and classroom with excitement. He was a tall, spare, handsome, quizzical man with a long nose, sensitive mouth, a deeply cleft chin, and a shock of receding hair which usually stood up like a cock's comb on the top of his head. Hooker was so full of enthusiasm for his subject that this enthusiasm radiated from him and infected all who were near. He liked to present a new idea in botany, make some shocking statement, toss it to his students like a bone to dogs, and then, leaning back against his table, arms folded, a half smile on his mouth, let them argue it out and in the end come around to his point of view. Sessions with Hooker were often explosively exciting. Not only medical students from the university, for whom the lectures were primarily intended, came to listen to him, but men from London and Edinburgh. So did the young assistant from the Botanical Gardens. He could not stay away.

David Douglas, his scrubbed, freckled face shining in anticipation, came daily to the hall to listen to his idol. Eventually he summoned the courage to introduce himself. The acquaintance became an intimate friendship which lasted until the untimely death of Douglas. It opened the door to the world for young David, to whom paradise itself was offered when Dr. Hooker invited him to go along on botanical excursions into the Scottish highlands.

As David Douglas became an expert botanist in his own right, he was introduced to the Royal Horticultural Society by Dr. Hooker himself. The society, the leading institution in the discovery and study of natural history, numbered among its members certain kings and noblemen, as well as most of the English scientists of the day. The society published elegantly illustrated and hand-colored works of natural history. It operated Kew Gardens; in fact, the Royal Society and the Royal Gardens at Kew eventually became one and the same. A young man who longed to go out and explore for plants would almost certainly need the sanction, protection, and funds of the society to do it at all. David had never dared to hope for such a thing, but it had happened—Dr. Hooker proposed it, and the society, respecting the judgment of one of its most eminent members, in 1823 engaged David Douglas to collect plants.

When he came to London and saw the splendid gardens with the exotic vegetation at Kew, he became full of a quivering excitement at what might be his first assignment. Would it be to Africa, the Orient, Brazil, possibly to the jungles of the Amazon, or to the South Sea Islands? The whole world at last lay before him. He was strong and healthy; it was only his eyes which bothered him at times, but he did not let them annoy him too much. He only had to cut short his reading. China . . . the Andes . . . India . . . the Sandwich Islands . . . Zanzibar . . . the dark Zambezi . . . ?

They sent him to New York.

The news was a blow. With a jolt he came back from the crocodiles along the Nile and orchids beside the Amazon. The United States. Well, it was better than nothing. It was a beginning. It was better than staying forever in Scotland or in London. So many expeditions had been made in America, however, both by Europeans and by the Americans themselves, that he doubted there was very much left for him to find in the United States. Especially in New York.

But Douglas's first mission was not to explore. He was being sent by the Royal Society specifically to examine American fruit-growing on Long Island and beside the Hudson, find out what was being done to develop varieties peculiar to America, and bring back such new trees as he considered would be likely to grow in England. To keep him happy, however, he was given permission and the funds to collect as many wild plants and seeds as he liked, and to go as far north and west as Amherstburg, Ontario, not far from Detroit.

On the whole, the journey was a good deal better than first had been anticipated. With letters of introduction from Dr. Hooker he found doors opened to him in pleasant hospitality. He was taken to see Bartram's garden—William Bartram had just died, but a niece showed Douglas around. He was entertained by Dr. John Torrey and Dr. Hosack. Exploring along the Hudson, where the Dutch had fine fruit farms, he also discovered a delightful sphagnum bog which contained Orontium, or golden club, and many pitcher plants and sundews. Sightseeing, he took his trip to Lake Ontario, and returned by way of Niagara Falls, admiring the falls as well as the plants which he obtained alongside. On his journey he collected a great many specimens. Then, arranging with several nurseries to ship trees to England, he left for home in December. His taste of America had been very rewarding after all.

His brief glimpse of the New World left him hungry for more: to go to the Far North and to the Northwest, to the land of Lewis and Clark, of Menzies and Mackenzie and Vancouver, of Franklin and Richardson and Herne. Savoring the unknown riches waiting for him, he often lay happily awake at night. He was sure that eventually William Hooker, in preparing his tremendous book called *Flora Boreali-Americana* from plants brought to him by those other explorers, would one day need a man to go out there specifically to collect for him and for the society as well, to fill the great gaps of botanical knowledge which still existed in the wilderness of America.

At last the Royal Horticultural Society, no doubt urged by Hooker, sent David Douglas on that very mission to the northwestern part of America. He was to be under the protection of the Hudson's Bay Company which, as much as possible, would shelter him and keep him safe from the Indians and other dangers. The society, however, hoped that he realized that he was not going into as easy and as social a situation as that which he had had in New York and eastern Canada. Food, living conditions, hardship—nothing would be easy. But it was exactly what he wanted. If he required ease, he could stay at home and walk in Kew Gardens. In order to discover new plants, he needed to go to the harshest and most demanding wilderness. He had no illusions about it, but Douglas knew he could endure anything for flowers and trees and birds, especially if they were new.

On a warm July day in 1824, he sailed on the Hudson's Bay Company's ship, the *William and Ann*. Prior to sailing he had had two splendid experiences with men who had been out there before him. He had visited old Dr. Archibald Menzies in London and examined his collections obtained from 1791 to 1795. He had gone to Dr. John Richardson's house and talked with him about his own explorations more recently experienced in the Arctic, in which Richardson and the others had nearly perished. They spoke of the forthcoming expedition which even then was planned by Franklin and Richardson to go again into the Arctic. Douglas and the doctor made a pact to meet, if fate should permit it, somewhere in that vast, cold wilderness of Canada.

When Douglas remembered the fearsome experiences which Richardson and Franklin had endured,* he wondered that any

* See Chapter 16, *Men, Birds and Adventure*, by Virginia S. Eifert.

man could wish to go back for more. But Dr. Richardson explained it simply: during the terrible retreat across the Barren Grounds, when several men had died and all had expected to be lost, Richardson had had to abandon most of his cherished collections. There had been no other course. Without any food but reindeer lichens and old bones scavenged from where the wolves had gnawed them, the men had become too weak to do more than drag themselves on and on, and not even the dedicated Richardson had had the strength to carry his specimens. He was compelled to go back to the Arctic to re-collect, and to find even more.

When Douglas at last was aboard the *William and Ann* and was suddenly feeling very much alone, he discovered a final kindness from Hooker. The doctor was sending along young Dr. John Scouler, a kindred botanist and naturalist, to go out to the Columbia River with him. As the sails billowed out to the wind and the ship set off, both men may have briefly recalled that, of the last three collectors sent out by the society, only one had returned alive.

After the long voyage around South America, and as soon as the ship finally dropped anchor at Fort Astoria, the new adventures began. Dr. John McLaughlin, the chief factor at Fort Vancouver, came down to greet him; together they went by canoe to the fort. From here Douglas would start into the interior, while Scouler stayed along the coast to collect.

With the tremendous snow-covered mountains on the skyline to the east, in a dramatic landscape whose like he had never seen before, David Douglas started out on six months of exploring and collecting along the Columbia River from its mouth to Celilo Falls. He traveled up the Multnomah River and inland from there; he collected furiously and with single-minded concentration. He wanted to have as many specimens as possible ready to send back with Scouler and the *William and Ann* when it returned from a more northern trip. Douglas shot birds, caught fish for food, slept under trees, and was alone a good part of the time. He was often entertained most cordially by the Clatsop Indians, who greatly admired his ability to shoot with a gun. This ability no doubt accorded him a definite respect by the Indians, who would have thought twice before risking an attack upon him while he was armed. They had a name for him—Man of Grass—because he was continually going about picking up plants. To most of the Indians, all plants were grass. If the Clatsops thought him a little bit crazy, they nevertheless had to respect his marksmanship.

Collecting was a joy, but it was hard. He had not known he could grow so desperately tired in only doing what he most liked to do, could be so hungry and so uncomfortable when he was at the same time so happy. The rains plagued him. Rainfall in that area was apparently endless. Day and night everything was damp, yet he knew that it was this same incessant precipitation which gave the land that extraordinary vegetation, that lushness of growth, those tall trees, those fountains of ferns, and those incredible, ankle-deep mosses. The moisture, however, caused him to lose many of his plant and bird specimens. When he could not keep them dry, they spoiled quickly. It took him an hour to dry his blanket before he could lie down wearily to sleep. Douglas reached his lowest ebb of endurance one day when he had become so fatigued that he could not stay awake while cooking his meager supper in the forest.

The exhausted botanist had put two small partridges to stew in his only cooking pot, and had hung it over the fire. Then, quite worn out, his tousled red head dropped forward and he went to sleep. Unknowingly, he gently eased himself into a more comfortable position and slept soundly until morning. And when he woke, he discovered that his meager dinner had long since burned black and the pot was quite ruined, with several holes burnt in the bottom. He was more hungry than ever, had nothing whatever to eat, and no pot to cook it in if he found something. But he did manage to make some tea by scouring the lid of his tinderbox in which to boil a little water.

When he left the rainy lands and went over the mountains to arid sand country of the interior, he suffered again. His eyes were now bothering him a great deal. It seemed that there was always a film laid over them nowadays which no amount of rubbing would take away. Sunlight made his eyes ache and burn. Although he could not always read very well, he still had no difficulty whatever in seeing a new flower when one was near.

In spite of the great heat and dryness of the interior—at least his specimens dried nicely here—he found the plant which had been named for the explorer, William Clark, who, with Meriwether Lewis, had come here in 1805. Douglas was now finding several species of the charming, ruffled flowers of Clarkia, along with the extraordinary little Lewisia in barren, open areas in the mountains. He was the first to bring out seeds of both, for propagation in England.

During those six months he collected 499 species of plants. Among them were some beautiful lupines and penstemons, an evening primrose of great beauty, several colorful monkey flowers, the starry Mentzelia, the Brodiaea, a new Collinsia, a lovely Spiraea, the Madroña, an Oregon grape, some new honeysuckles, and a great many others. He had come upon a splendid treasure.

On this journey he had finally managed to determine the height of the new fir, the one which would be called Douglas fir, or Douglas tree, since it was neither a true fir nor a true spruce. Now as he measured a fallen one he found that it was 227 feet long and forty-eight feet in circumference at a point three feet above the ground. But he could not manage to get a cone with seeds. The fallen trees did not have them. On the living trees the cones hung immensely high in the tapering upper boughs. The structure of the cone amazed him. It was loose, like a fir cone, but hung below the twig as the spruce cones did. Springing from the bases of the scales extended slender, three-pronged projections or tags which gave the cone a shaggy look. It and the tree were certainly unique. He could anticipate the delight which Hooker would feel at this discovery.

All the western conifers thrilled young Douglas. He was in a realm where perhaps more of the truly elegant and beautiful coniferous trees of North America were to be found, few of which had as yet been discovered or named. His finds included the Sitka spruce, the Coulter pine, western white pine, ponderosa pine, silver fir, Sabine pine, the noble fir, and the lovely fir. Any suggestion that a new one might be found sent him on a fast journey into the wilderness in the hope of finding it.

Thus, one day in a dirty Calapooie Indian village when he was offered some large, nutlike pine seeds to eat, he became so excited by their size that he lost interest in food. Here indeed was something new and remarkable. He tried to persuade the Indians to tell him where they had found these seeds, but they could only vaguely wave their hands to the south, saying the trees grew there. He alerted trappers and hunters to look for them, was anxiously on the alert himself. For he wondered at a pine whose seeds were big enough to be eaten as nuts; wondered how large the cones must be to contain such seeds, and how great were the trees themselves. The Indians in describing the cones had held their hands apart for an incredible distance to indicate their size. Allowing for some exaggeration, the cones must be enormous.

He came back to Fort Astoria without finding the mysterious pine, but in time to contact the waiting *William and Ann,* now preparing to leave for England. He saw to the loading of sixteen large boxes of plants collected on the Columbia and eight more packages of materials collected inland. Included was a well-packed box of bird and mammal skins and a parcel of Indian artifacts and crafts. But, in a chest by themselves, safe from sea water and from heat, were over a hundred kinds of named and unnamed seeds of precious trees and flowers of the Northwest. He knew, as everyone else did, including the impatient Dr. Hooker, that sailing ships did not always reach their destinations. His precious cargo of plants, seeds, and specimens might very easily be lost en route. But with samples of all of them kept in his possession, to be brought by him personally on his eventual return, perhaps his discoveries would one day reach England, for the enrichment of the world.

The race to collect was over for a time. The ship was gone, his specimens were gone, for good or ill, his friend Scouler was gone. The growing season was over, and he was desperately tired. He was suffering from an infected knee incurred when he fell on a nail; he needed to stay off his feet. The winter of long rains which set in gave him rheumatism. At Fort Vancouver, therefore, he settled down to classify, mount, and preserve more specimens, to collect mosses and wood samples when the weather and his knee would permit. At the same time, he let himself sink very low in spirits; the rains and fogs oppressed him. He was lonely. "In all probability," he wrote mournfully, "if a change does not take place, I will shortly be consigned to the tomb."

As soon as the weather improved, his knee was well, and plants began to blossom, David Douglas cheered up. He felt so much better that he decided against going back to England in the spring. He felt he really had not done justice to the Northwest and simply must stay another year. He would then hike across the continent to Hudson Bay, there to board a ship for home.

With this extension of time, he could accomplish a great deal. He only wished he could see a little better—the dimness was more than ever veiling his eyes, and it had somewhat impaired his shooting ability. Realizing that it would be wiser to go home where something might possibly be done to help him, he nonetheless remained. As long as he could see at all, he would stay in America to collect. Once in England, he might never get back to America again.

Springtime, and Douglas set off with a company of the Hudson Bay express going in canoes up the Columbia. He would go part of the way with the men for protection through the Indian country around Celilo Falls, where the natives were troublesome. The journey was slow, which was what a botanist preferred. While the men pulled the boats or paddled when they could, Douglas ranged the rocky shores or scrambled up the slopes of mountains. He found there was still snow in the high places; when the express reached the upper portions of the river, the drifts were too deep for botanizing.

At the junction of the Spokane River, the men went one way and Douglas went another. Next spring he would go all the way with them, but just now he entrusted to the express a box of seeds and plants to be taken to Hudson Bay. At this parting point, they had come six hundred miles—six hundred glorious, rugged, and indescribably magnificent miles. Now he was alone in the echoing quiet of the high mountain snows. Where the snow had melted, the alpine meadows were golden with the dancing bonnets of the glacier lilies. He was alone—but in what a splendid company.

Determined to find the mysterious pine with the large seeds, he sought out an Indian village to secure a guide. The trees, the

Calapooies had said, grew in the Blue Mountains to the south, but most Indians seemed to be afraid to go there. The Man of Grass, they said, could go and visit the dread place if he wished, but he would have to go by himself. They feared the evil spirits in those wild and desolately remote parts; besides, there were tribes living there who were hostile. They themselves did not go near, but traded with other, braver Indians for the delicious pine seeds.

Baleful spirits and hostile Indians did not alarm him. Obsessed with finding that unknown tree, he went back to Fort Vancouver to obtain supplies, and on September 20 he set out to the south in company with thirty men who were bound for a distant fur post. Part of the way they traveled by canoe, then with horses, then moved slowly southward on foot or on horseback. It was a tedious and difficult journey. The country was so mountainous and rough that only a few miles were covered between sunrise and sundown each day. The horses were slow, and at the night's camp they often strayed and caused further delay until they were rounded up the next morning. The heavily forested country was much entangled with underbrush, the way hampered by huge fallen trees, the trail, such as it was, lying concealed beneath the profuse growth of ferns and salmonberry bushes. The horses stumbled and fell and their packs came off. Men with axes often had to go ahead to clear a way. And it rained and it rained and it rained. David Douglas had never known such precipitation as he encountered here in the Northwest Coast country, nor ever had a harder time trying to keep himself and his specimens dry.

They were in the country of the grizzly bear. None had been seen until one of the hunters, out to find meat for dinner, was leaped upon and attacked by a great beast which he had startled as it was tearing open logs in a burned area. The bear had charged in enormous fury and, before the hunter had managed to scamper up a tree, had ripped off most of his clothes and a good deal of skin, too. The man was fortunate to be alive. Douglas, knowing that he would have to be more watchful when he was poking about in the tangles of shrubbery and ferns, kept his gun with him at all times.

Not bears, however, but a deer almost caused his end. Pursuing a large black-tailed buck, he went tumbling head over heels into a ravine full of fallen trees and tangles of driftwood swept there by past freshets. The fall knocked him unconscious and he lay there out of sight, bleeding from a ragged leg wound made by a tree snag.

Since he was often away from the group for hours, the hunters evidently did not start out to look for him, but if they had—and perhaps some of them finally did—they would have had no inkling as to where he was in all that jungly wilderness. Unconscious and undiscovered, he lay there for five hours. As he was beginning to stir a bit, groaning, some Calapooie Indians, out hunting for chinquapin nuts, heard him, peered gingerly into the ravine, and found him lying there. They helped him out and found his horse, but he could not mount. Douglas's chest pained him terribly. He could neither climb on nor stand straight. While one of the kindly Indians led his horse, Douglas, like an old man hunched over a stick, limped back to the camp.

His pain was extreme and there was no doctor to help him. In desperation he followed a practice of the times—bleeding a person in pain was expected to help him greatly. So he bled himself in the foot, and then rolled in his blanket to try to sleep. Next morning he bathed in the icy water of the river. After this drastic treatment he found that he could actually draw a deep breath without its hurting too much, and felt well enough to travel. Parting company with the fur men, he went on with one of the Indians who had helped him from the ravine. He was determined now to find the elusive tree.

They went on together for several days without finding what he sought. At one of the camps when he had left the Indian behind to attend to the careful drying of some specimens, and had gone out alone to look for the pine, he had found, not the pine, but another Indian. The man was evidently as much startled as he and immediately drew his bow. Douglas, his heart pounding, but seeing how obviously frightened the stranger was, boldly put down his gun and gestured to the Indian to do the same with his weapons. With some hesitation, the man did so. Douglas gave him some tobacco; the Indian relaxed. It had been a close thing, but it was over; he was now more concerned with what the Indian might tell him about pines.

Taking out a pencil and his notebook, he made a sketch of the cone as it had been described to him, and the Indian nodded, smiling, and pointed to mountains not far distant in the south. His new friend offered to go with him and show the way.

That was the day, October 25, 1826, when the western sugar pine, tallest of all pines, was discovered in central Oregon by a

Western Sugar Pine

white man and brought out to civilization. Douglas stood at last beneath trees whose huge boles and high-held boughs with their enormous pendant cones embodied western grandeur itself. He could scarcely comprehend the size—trees came large in this country, and he had already seen some huge ones. But a sugar pine which he found blown down in a recent storm was measured at 215 feet, and three feet above the roots its circumference was fifty-seven feet and five inches, while 134 feet above the ground it was seventeen feet and five inches around. From cracks in the smooth brown bark there oozed a sweet, clear resin which would

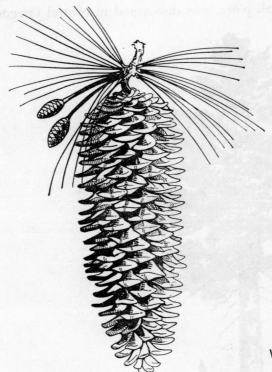

Western Sugar Pine

give it the name of sugar pine. He could not, of course, reach the cones, but he could see them up there hanging like purple-brown sugar loaves from the pendulous tips of the branches. He had not really visualized them like that, like enormous ornaments so very far out of reach. There were some on the ground, but they lacked seeds; he must obtain fresh cones and fresh seeds if he was to bring this new wonder back to England.

The trees were much too vast to be climbed. Few had any branches lower than eighty feet up. He therefore resorted to a method he had often used to get something he needed from on high—his eyes were still reliable enough to draw a bead on a pine cone a hundred feet in the air and clip off the twig just above to bring it down. He thus obtained three excellent specimens. They were the biggest pine cones of any tree on earth—twenty inches long, madder-purple inside the scales, a rich brown outside, with seeds as large as a grain of corn. Admiring his find, he suddenly realized that he was not alone.

At the sound of his shooting, eight Indians had appeared. They were painted with red earth and had bows and arrows, wicked-looking spears, and unfriendly expressions on their faces. One man, with a scowl, was very ostentatiously sharpening his flint knife with an ominous *fweet-fweet-fweet* which grated on the nerves and sent a chill up Douglas's spine. He gestured, used what Chinook and Umpqua he knew, trying to explain what he was doing. He offered them smokes, and they seemed momentarily pacified, until the one resumed whetting his knife. Hostile once more, the others drew closer to the man with the big cones.

The red-haired botanist stepped back half a dozen paces and put down his cones. He cocked his gun. He pulled out one of his pistols with his left hand, and then stood belligerently, a two-gun man well armed and ready to fight. He knew he was outnumbered, but he stood thus defiantly for what seemed like hours. The Indians stared at the white man; the botanist, squinting in the light, returned the stare. The Indians relaxed. The knife-sharpener put away his weapon and his whetstone. The leader amiably told Douglas they would let him go if he gave them some tobacco. But he was so far from being intimidated that he sturdily returned the request for fair exchange—he would give them tobacco if they would collect some cones for him. While the Indians scattered on their botanical mission, he picked up the fresh cones and hastily strode back to his camp. In fear that he might betray him, Douglas dismissed the Indian guide.

And there, all alone in a dangerous wilderness, with temperamental Indians not far away, David Douglas sat down by his campfire and looked with immense satisfaction and pleasure at the long-sought treasure. What a sensation these cones would make in England. What a thrill for William Hooker as well as for all the botanists at the Royal Society and the gardens at Kew, who never, never in all their lives, had seen cones like these. He would name the tree for one of them, Dr. A. B. Lambert, a much-admired botanist, and the name he proposed held. The western sugar pine became *Pinus Lambertiana.*

The long trip back to Fort Vancouver was as severe as all the preceding experiences. For protection, he joined a returning fur party and then, because of an impending Indian attack, had to spend a fearful night hidden in the bushes. In a driving rain shortly before morning, the Indians came, fifteen of them, but were scattered by men of the fur party who fired blank shots at them.

Proceeding wearily in the rain toward the fort at last, the men had to swim their horses across the Santiam River, and, in the rushing tumult of the waters, one of Douglas's pack saddles was swept away and he lost almost all of the specimens which he had collected so laboriously on the two months' excursion. In the remaining saddle bag, however, carefully wrapped, reposed the big pine cones and other valuable specimens. He at least had these and, in the light of their discovery and importance, the losses and the anguish surely were eminently worth enduring.

After another long winter at the fort, David Douglas packed all his specimens and seeds for the long journey back to England. Part he would send by ship, the remainder he would take with him in March when he set off on foot for Hudson Bay.

The long journey began on March 20, 1827, when the express set off up the Columbia with two heavy boats loaded with bales of furs and led by Edward Ermatinger, clerk of the Hudson's Bay Company. Douglas walked most of the way. On foot he could go faster than the boats and could see and collect more. Up the Columbia they went, up to the snow fields where again he reveled in the sight of the lovely golden glacier lilies blossoming along the edges of the snow. Even though it was the wrong season to transplant them, he felt impelled to dig up some of the bulbs and take

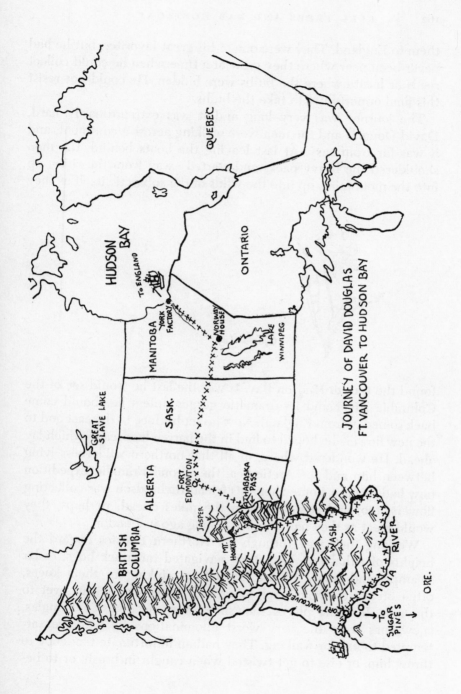

HUDSON BAY

QUEBEC

ONTARIO

To ENGLAND

YORK FACTORY

NORWAY HOUSE

MANITOBA

LAKE WINNIPEG

GREAT SLAVE LAKE

SASK.

ALBERTA

FORT EDMONTON

ATHABASKA PASS

JASPER

MT. HOOKER

BRITISH COLUMBIA

WASH.

FORT VANCOUVER

COLUMBIA RIVER

To SUGAR PINES

ORE.

JOURNEY OF DAVID DOUGLAS
FT. VANCOUVER TO HUDSON BAY

them to England. They were one of his great favorites, but he had never been near where they grew at a time when he could collect seeds or locate where the bulbs were hidden. He could not resist this final opportunity to take the bulbs.

The journey was very long and it was extraordinarily hard. David Douglas and the men were walking across a continent, and it was far from easy. At last leaving the boats behind, the men shouldered the heavy packs and started away from the river, up into the mountains, up into the wilds of the subarctic itself as they

Glacier Lily

found the trail to Hudson Bay. It was the last he would see of the Columbia River and its dramatic gorges unless he should come back someday from England. Now he set his face to the east and to the new flowers he hoped to find in the barren landscape which lay ahead. He wondered where, in all that northern wilderness lying between him and the Arctic Sea, the second Franklin expedition now had its base. Somewhere his friend Richardson was collecting flowers, too, and somewhere in the miles ahead, perhaps, they would meet as they had planned it long ago in London.

Working their way through the Northern Rockies toward the heights of Athabaska Pass, they navigated tamarack bogs, cedar swamps, and soggy tundra, sometimes sinking to their knees, often breaking through softening ice, frequently getting wet to the skin. In some places snowshoes were needed. To Douglas, snowshoes were the very worst abominations he had ever attempted to use for walking. They had an unfortunate tendency to throw him, or else to get twisted when caught in brush, or to become unlaced. They would drop him suddenly all in a heap into deep, wet snow. Yet, hard as they were to use, they were often the only means of getting over the soggy drifts.

Nine men on snowshoes, men bent with heavy back packs, were, as he described it, on "a journey over such an inhospitable country . . . one falling, a second helping him up, a third lagging and far behind, a fourth resting smoking his pipe, and so on." They walked gravel bars of rivers, forded one winding river fourteen times, endured piercing cold. Still farther northward, they found the snow four to seven feet deep, with not enough crust to hold them up, and had to resume the execrable snowshoes again. At night they sometimes slept in a snow pit for warmth, sometimes on beds of balsam boughs. One morning the campfire, kindled on the surface of the snow, had sunk itself six feet into the drift, making what he termed "a natural kitchen."

The great, cold, slate-purple mountains and their snows . . . up Athabaska Pass . . . a little side trip for Douglas, seeking the high-altitude flowers . . . returning after naming some peaks for botanists of his acquaintance—Mount Hooker, Mount Brown, Mount Fraser . . . a scarcity of food . . . these were all part of the explorer's life. When Ermatinger one day shot a partridge for dinner, Douglas managed to skin it before it was cut up, and eventually conveyed it all the way to England. Here it was later named Franklin's grouse in honor of Sir John Franklin who, with Richardson, had first found the bird in the wilds of the Arctic. These birds had many times helped to save that hard-beset expedition from absolute starvation. The grouse did it again on the 1827 journey of the Hudson Bay express.

In the heights of what is now Jasper National Park, the men were met by a packer and guide of the Hudson's Bay Company, who had brought horses as part of a prearranged, continent-wide plan operated by the efficient company. The party thereupon proceeded thankfully on horseback—all but the botanist, who still stubbornly preferred to walk. At Jasper House, a station of the company on Brulé Lake, they rested, then proceeded 184 miles to Fort Assiniboine in three days. Since this was dull country in which there were no new plants, and few of the familiar ones themselves, the botanist was willing to cover the ground as fast as possible.

He was, besides, growing impatient to reach Fort Edmonton. While the men went more slowly in order to shoot food, Douglas went ahead and reached the fort before them. His great haste was to check on the box of seeds and specimens which he had forwarded earlier on its way to Hudson Bay and, he hopefully expected, it should be waiting for him at Fort Edmonton. He was uneasy about it, wondering if the collections had been damaged, fearful that they might not have arrived at all.

The box was there. Douglas, up at dawn to examine the precious seeds and pressed plants, was greatly relieved to find them in such excellent condition. He learned with both gratitude and dismay that Thomas Drummond, botanist and naturalist who was with Franklin and Richardson in the Arctic, had visited Fort Edmonton and had looked at Douglas's collections. Seeing that they were damp, he had kindly changed the papers on the pressed specimens and carefully dried the seeds.

Although Douglas truly appreciated this gesture, he would frankly have preferred that the noted Dr. Drummond had not been so diligent in examining another's rare new species. In Douglas's journal it is obvious that among botanists honor is not always triumphant over avarice, in acquiring another's discoveries and taking credit for them. Drummond, in fact, could have helped himself to something special from David's collections, and if he, Drummond, had succeeded in getting them back to England first, he would thus have the right to claim them as his own. Douglas would like to have trusted Drummond, and had no real basis for any such suspicion, but, after all, one never knew. . . .

At Norway House, the next stop, mail awaited. He was relieved to learn that plants which he had sent by ship from the Columbia had indeed arrived in England in good condition. There was also a letter informing him that his father had died. At Norway House he took leave of the men of the express. At the same time, he was delighted to find that, on the morning he expected to leave, three great naturalists and Englishmen had just come in from the Arctic regions—the estimable Dr. Drummond himself, Dr. John Richardson, and Sir John Franklin. They and their men had just returned from their second expedition to the Polar Sea and Douglas was relieved to learn that this had been a much kinder trip than the first one, was gratified to have the opportunity to examine Richardson's new collections, see those which Drummond had acquired, and compare notes. Considering how far all of them had come, theirs was really a rather momentous and wonderful meeting in the wild lands of upper Canada. When Douglas and Richardson had planned it, long ago in the comfort of a London town house, it had seemed an almost remote possibility that they should actually meet in America.

Douglas sailed from York Factory on Hudson Bay on August 20 and reached England to find that he was something of a celebrity. London, which adored explorers, greeted him with acclaim. He was made a member of scientific societies, he gave lectures, he wrote papers, he was admired. Douglas's modesty was almost lost in this heady atmosphere. Dr. Hooker's honest happiness, as well

Mariposa Lily

as his pointed dissatisfaction, over all he had brought back, however, was a leveling influence. No matter how many fine things David discovered, Hooker would always wonder why he had not brought more.

But Douglas had sampled wilderness, and civilization could never again keep him very long. Being a social lion soon palled. He longed to go back to the Columbia, to go to other places, to collect again. Eventually, through Hooker's efforts, David Douglas once more was on his way to adventure. In spite of his failing eyesight, he ranged up to Alaska. Then he went down to California, where he saw the giant redwood trees and collected beautiful new wild flowers which he felt would do well in gardens—gilias and California poppies, baby blue-eyes, cream-cups, mariposas, and dozens more.

He was, besides, probably the first man to discover gold in California. Clinging to the roots of some of the plants which he sent to England from northern California were grains of actual, genuine gold. Treasure lay in the Royal Society's herbarium for seventeen years before it was officially discovered at Sutter's Mill.

Now at last on his way back to England with his new collections, Douglas went by way of Hawaii—the customary route for many of the ships before heading down to the tip of South America. Botanizing in the lush jungles of the islands was an incomparable experience. He climbed mountains, looked into volcanoes, found beautiful tropical trees, new ferns, new flowers. He would have to come back when there was more time . . . he must surely come back.

It was, however, growing more and more difficult for him to see. One eye was now totally blind, and the other was always inflamed and painful, sometimes exuding blood after excessive exposure to light. There was always a film over it, and he would not let himself think of the day, perhaps so soon at hand, when he would be quite blind and shut away forever from his lovely wilderness. A London oculist, before he had departed, had fitted him with the strongest possible eyeglasses, but they had helped him very little, if at all.

David Douglas, however, could still climb mountains. With an amiable little black spaniel for company and to assist him in finding the way, he managed to climb Kilauea and Mauna Loa. As he watched the lava flowing down the mountainside, the fumes of the volcanoes hurt his eyes.

Douglas intended going to England from Honolulu on the next ship. However, since there was no prospect of a ship's arrival for several more weeks, he had time for still another expedition. It was as if he tried to crowd all he could into the remaining days during which he could see, packing away memories for the interminable days of darkness. In July, 1834, therefore, he went in a small boat to Hilo and visited friends at the mission. With the dog leading the way, he took a path up into the mountains.

Before he started on the trail, his worried hosts at the mission warned him to watch out for the bullock pits, certain deep holes which had been dug by the natives, then covered loosely with matting and brush, into which wild cattle fell and were captured. Douglas promised to be careful, but, in letting him go off alone, it is doubtful that anyone knew the extent of his near-blindness.

The pits were not easily detected on the trail. With his eyes so bad, he had to be slow and very cautious—the presence of the dog was an assurance which he appreciated. He paused at the first pit and saw a bull in it. The second pit was empty, and the third . . . the third . . .

Several hours later when natives came up past the cattle traps, they saw a small dog waiting beside the third pit. Below was a black bull, and under the bull's restless feet, trampled into the earth, lay a battered tin botanical collecting case and what was left of a red-haired man. He had been gored and mangled. David Douglas, botanist, had reached the end of his trail of adventure.

It was a tragic end, but perhaps it was mercifully swifter than the end might have been for him when he inevitably lost his sight. In his brief thirty-five years he had brought back to England and the civilized world more kinds of plants than any other man of his time. By means of seeds which he collected, he had introduced into gardens over the world some of the most splendid trees and flowers of America. And, strangely enough, descendants of many of those lovely western plants, propagated by nurserymen and seedsmen in Europe, eventually came as improved seeds and roots to the plant companies of the United States. In a remarkably roundabout way, some of North America's own wild flowers returned as uncommonly beautiful, cultivated garden plants. Among these were the Clarkia, Virginia clematis, Broadaeia, Gilia, Mariposa lily, Oregon grape, California poppy, Fritillaria, Gaillardia, the annual sunflower, coral bells, tidy-tips, California trout lily, blue flax, a dozen lupines, several brilliant monkey flowers, Nemophila, eleven kinds of evening primrose, America's only wild peony, called Lenten rose, eighteen elegant penstemons, colorful phloxes, the golden currant, the fuchsia-flowered gooseberry, and the bitterroot or Lewisia. They were his legacy to the land he had learned to love, to which he had come as a stranger and from which he had carried treasure which would be sent back one day with interest.

*Colorado
Columbine*

10. The Plant Men

As the sixteenth and seventeenth centuries were the age of the herbalists, the eighteenth and nineteenth centuries were the age of the botanists. This was the period in which men of many nations, including North America, were tremendously eager to know the secrets which the New World held in its flora and fauna. It was the period of the collectors, the heyday of the explorers, who were searching in the name of science. It was their treasures, often painfully sought out and brought back to civilization for analysis, study, and classification, which provided for the stay-at-home scientists some of the great experiences in the expansion of human knowledge during those centuries and a better understanding of the natural history of North America.

Ever since the days of John White, Thomas Hariot, John Smith, and John Josselyn, collectors had been taking to Europe the first choice discoveries of North American plants. A century and a half after Roanoke and Jamestown, it was the Americans themselves, or those who had adopted America for permanent or long-term homes, who were scouring the woods and marshes for plants. If these men were sponsored by a scientist on the eastern coast or in

Europe, they were usually asked to collect several dozen or several hundred good specimens of each species. The collectors were to press them carefully, identify their locality, and bring back as many complete sets representative of certain areas as they possibly could. At that time, plants were used as a friendly medium of exchange between botanists in America and Europe, as well as an actual financial exchange. If a botanist in Europe was willing to pay for a specimen or a full set of specimens, a collector might often finance the cost of his entire expedition, or at least defray the expenses of his sponsor or the group with whom he traveled. Collections in the nineteenth century became a mild form of speculation.

If a botanist had kindly been given some pressed plants by one of his scientific correspondents, perhaps from across the sea, it was the usual thing to reciprocate in kind, and hope that what one sent in return was just a little bit better or more rare than the other's. The size of one's collection at that time was a status symbol. Like acquiring stamps or matchbooks or African artifacts, the more specimens one had, the better were his state of acquisitive happiness and his feeling of success in the world of science. This often induced a greater urgency than mere scientific curiosity in obtaining new plants and getting them properly classified. Competition among botanists was a great spur to the exploration of America and the discovery of the American plants.

More and more it was realized that this indeed was a very large country—it was big geographically as well as botanically, and much of its landscape and history were determined and identified by the plants which grew there. Economically, vegetation might also determine which settlements might be built in certain areas. Immigrants from Northern Europe and Scandinavia were more likely to choose as settlements the areas in the Northern United States where they would find a familiar landscape of spruce, fir, pine, and peat bog, while Spaniards were apt to choose the more tropical and familiar areas of the South and Southwest where palms, agaves, and other exotics grew. The vegetation itself largely determined the crops which might be expected to grow there. Forest assets determined the location of mills and factories. Botany went hand in hand with the national expansion, with the national economy, and with the national growth of knowledge as well. Americans had to know their country and from what they sprang, must know where they had been before they could know where they were going. Knowledge of natural resources and natural history was one of the ways to know America.

In the nineteenth century, the major botanists were developing in the East, where the culture was. New York State, then Philadelphia, then Washington, D.C., became centers of science and, since botany meant not only plants but men, it was these men of this period who largely built the sum of today's immense botanical knowledge. They were the ones who not only collected on occasion, or who awaited the collections of other men, but who explored and made the botanical background which is taken for granted today in such books as Gray's infinitely detailed *Manual of Botany;* in Britton and Brown's *Illustrated Flora of the Northeastern States and Canada,* and its massive revision undertaken a few years ago by Henry A. Gleason; in Mathews' *Field Book of American Wild Flowers;* in Sargent's *North American Trees;* in McIlvaine and McAdam's *One Thousand American Fungi;* and in innumerable other books which, through the years, have been the standard works of identification of plants in America. For each plant described in these books, one man or many men must have had to discover, study, and evaluate in order to distinguish that plant from all others, had to find out its life history and learn of its range.

Through the years there have been ponderous reference works of botany intended for botanists, and there have been a great many well-illustrated books of wild flowers, trees, fungi, and other plants intended for amateurs who simply want to know about their surroundings. In these books the meat and marrow of the scientific discoveries have been rendered into something intelligible to those without an extensive botanical background. As keys to the doorways to the world, the handbooks often lead disciples into the more concentrated, complicated, and dedicated fields of natural science. Like John Bartram in the classic tale, they must first really see their daisy before they can focus on the rest of the wild things.

This is the role of the popular plant books and other volumes on natural history which lead to that moment of awakening.

At first, in the growing flood of newly discovered plants—in bales of specimens to be examined and the unknowns puzzled out —there was little time to study plant mechanisms, heredity, and genetics, the complex chemical or sexual attributes of plants, or growth, photosynthesis, morphology, cytology, or plant pathology. With the increasingly extensive exploration of America, plants were being found in such great numbers and brought in for identification so rapidly that indeed nothing could be done but to handle the vital and immediate task of trying to classify them accurately, either by the old sexual system of Linnaeus, or by the newer and more usable natural system. The business at hand was taxonomy, of determining genera, species, and subspecies, and of amassing a growing quantity of names and descriptions for the botanical manuals and catalogues of America.

Compared with John Josselyn's little more than two hundred species in 1663, Michaux with his pioneering book called *Flora Boreali-Americana*, published in 1803, had greatly enlarged the scope by describing 1,613 flowering plants and ferns. Gray's first edition of his *Manual of Botany*, in 1848, listed 2,076 species for the eastern states. Eight years later it had grown to 2,426, and, by 1867, the fifth edition included 2,642 species. With the range extended to include the Great Plains to the 100th meridian, the sixth edition added 515 more, bringing the number to 3,157 species. By the time the eighth edition came out in 1950, the vascular plants named for the eastern half of the country had reached 8,340.

Throughout the story of American plants, as in other sciences and in history itself, there has been a continual linking of men, personalities, and discoveries. One had connection with the other, just as John Le Conte in the early years of the nineteenth century was linked with John Torrey, who was a contemporary of Asa Gray, who was the guiding light and link between men of the Victorian age and today. One botanist knew the others in his field, compared notes, perhaps had rivalries and often had violent disagreements and feuds, but while they disagreed and argued and studied, they were all at the same time working together for a mutual end—the search for truth.

When Dr. John Eatton Le Conte collected plants on Manhattan Island and, in 1810, published his *Catalog of the plants on the island of New York,* he was setting the scene for the plant books which were to follow. John Torrey, a friend of the Le Conte boys, was thirteen years old when that book came out; it spurred him to think about a book of his own. He helped Amos Eaton collect plants for the latter's *Manual of Botany* in 1817, when Eaton was in prison and could not get out to collect. Torrey struck up a correspondence with such notable botanists as Pursh and Muhlenbergh, Willdenow, and Persoon as well as most of the other men who were soon to leave their names on American plants.

Torrey became a physician in New York but, at the same time, he held fast to his botanical pursuits. He believed in a careful exploration and study of one's immediate surroundings; although distant expeditions all had their place, so also did regional botany, and to this end he traveled widely, like Thoreau, in his own area, and as far south as the Pine Barrens of New Jersey. He compiled a book which he called *A Catalog of the Plants growing within thirty miles of New York.* In this work he became so much better acquainted with other botanists that he began, at the same time, to see with some disillusionment that not all men of science were completely honest or always especially careful in their reports. He discovered how careless Frederick Pursh was at times, knew he had lifted first rights on *Drosera filiformis,* the thread-leaved sundew, which Constantine Rafinesque had obviously discovered first. Pursh also claimed discovery of some of the Lewis and Clark plants, but everyone knew that Pursh had never been west of the Mississippi.

That western country in the first part of the nineteenth century was full of hazards, not the least of which was the danger besetting scientific collections themselves, and the difficulties of bringing back pressed plants from an experience as rough and tremendous in scope as the Lewis and Clark expedition—about four thousand miles, one way, between the Mississippi River and the Pacific Ocean. It was somewhat miraculous, in fact, that any

specimens at all managed to arrive in the hands of the eager botanists in Philadelphia. Even after the plants had reached the comparative safety of Pennsylvania, more than the usual amount of
difficulty seemed to pursue the Lewis and Clark specimens.

A large packet of plants, all of those which had evidently been
laboriously collected by Meriwether Lewis, had been swept away
when baggage had fallen into the river during a boat upset on the
upper Missouri. The only record of these finds, therefore, was
to be seen in Lewis's journal, together with occasional small,
crude sketches which he included only too rarely in the text. However, during the greater leisure when the expedition wintered on
the Pacific Coast, Lewis collected a good many excellent specimens including the salal, Oregon grape, and a new fir. The latter
was the unknown tree which Menzies himself had found in 1792,
and which later would be called Douglas fir. As the men started
back up the Columbia and climbed over the mountains to the
headwaters of the Missouri, and proceeded down that river, Lewis
collected more plants. He ultimately managed to bring back 155
specimens. Most of these were totally new to science and many
of them were of unknown genera. He very much regretted those
which had been so inadvertently lost. Considering the hardships
of the trip, however, it really was a wonder that anything at all
was brought back.

Delivering the specimens in Philadelphia after the long journey * through the wilderness was one thing, but their fate after
that might be quite unpredictable and oftentimes disastrous. Dr.
Benjamin Smith Barton had promised Lewis and Clark to examine
and identify their plants, but when they at last were brought to
him, he turned the collection over to young Frederick Pursh, a
German botanist who was in the United States to study and collect
plants for a book which he contemplated. Barton made the young
man his protégé, and then did him the great favor of letting him
work over the expedition's plants. Perhaps if Dr. Barton had
known what was going to happen to those priceless specimens, he
might not have been so free with them.

Frederick Pursh took them with him to Germany—or at least he

* See Chapter 8, *Men, Birds and Adventure*, by Virginia S. Eifert.

took some of them; no one could be sure afterward; it was all very vague and astonishingly careless. The collection, at any rate, simply vanished. Lewis died; some said he had been murdered. Clark was very busy as Indian agent in St. Louis. The Lewis and Clark journals were edited by Nicholas Biddle, with the assistance of George Shannon, who had been on the expedition, and eventually were published in Philadelphia in 1814. That same year Pursh's book, *Flora Americae Septentrionalis,* in two volumes totaling 725 pages, was published in London, and in it Pursh included plants from that western expedition. Some of these, as Torrey and others recognized, he failed to credit to their discoverers, but gave himself the honor of having found them. This was an unfortunate habit indulged in by Pursh; some of his contemporaries criticized him severely for it.

It was thought that Pursh had probably left the Lewis and Clark collection with Dr. A. B. Lambert in London. In some way, no one knew how, although it was suspected that an unknown benefactor had purchased them from Lambert, some of the Lewis and Clark specimens finally came back to America and were given to the herbarium of the American Philosophical Society in Philadelphia.

But these were only a very few of the 155 specimens which Lewis had managed to convey with so much difficulty from the Far West. If anyone ever wondered where the rest of them were, no one did anything about the matter. Then, many years later, there took place one of those wonderfully gratifying discoveries which is as pleasant to think about now as it may have been then.

More than three quarters of a century after the return of Lewis and Clark, the cluttered quarters of the Academy of Natural Sciences in Philadelphia needed to make room for some improvements. In the process of digging out the scientific accumulations and rubbish of more than a century, someone came upon several very old, beetle-ridden bundles of pressed plants. The labels bore handwriting which was identified as that of Frederick Pursh. Notes appended showed that these indeed were the missing specimens of the Lewis and Clark expedition of 1803-06. For many decades they had been stored away at the mercy of insects and other deteriorating factors which had been at work. Some of the plants had been reduced to powder; some were in fragments and were difficult to identify. But, in general, in spite of the passage of time and lack of care, the rest were in a surprisingly good condition. They were remounted and preserved, lovingly enshrined in the Academy of Natural Sciences, where the Philosophical Society had apparently deposited them soon after Pursh had finished his work, and where they had been concealed ever since.

Botanists examining them found that neither Lewis, who had discovered it in the West, nor Pursh, who ought to have given it a name, had identified the Osage orange. Lewis had brought seeds to Philadelphia and these, planted and tended, had grown into trees. From them Constantine Rafinesque, some years later, identified the Osage orange and gave it the botanical name which it bears today, *Toxylon pomiferum.*

According to these once living records, Lewis and Clark also had discovered many plants which, for lack of names at the time, were credited to later explorers—or else to the overeager Pursh,

Clarkia

who was not averse to taking credit, on occasion, for something he had never seen. Another botanist, De Candolle, named one of Lewis's discoveries *Purshia tridentata.*

Pursh, however, honored the two explorers by naming two new genera of plants for them. On a mountainside in Montana, Lewis had come upon dancing hordes of flowers growing on two-foot stems, flowers of scarlet and crimson, of pink and purple, with four ruffled petals and an airy way about them which was delightful to see. This was named *Clarkia pulchella.* The little stout-rooted, pink-flowered plant, which the men had found in barren mountain country of Idaho and Montana and which the Indians called bitterroot and used for a richly nourishing food, was named *Lewisia rediviva.*

Barton, like Dr. Torrey, had his finger on the pulse of botany and they both were acquainted with the man whom Audubon had called an "odd fish," Constantine Rafinesque. In spite of the latter's eccentricities, he discovered a good many new plants and

Lewisia

left his name not only on some of these, but on new species of animals, especially among the fishes and mollusks.

Torrey evidently was one of the few men who liked Rafinesque and he held to that friendship when everyone else had turned from the "Mad Scientist," as he was jokingly called. For Rafinesque was a curious blend of deep scientific knowledge and extreme ego. He always had the wish to create a stir, to make a name, to cut a swath, to be noticed and admired. A small, dashing man, dark and handsome, with a mop of curly black hair, he had been deserted by his wife, perhaps for his very flightiness and insupportable genius. He had thus turned to concentrate entirely on botany and zoology as a substitute for that other portion of his life which was now closed. He was undeniably a remarkable man; it was his consummate conceit over the fact which disturbed his contemporaries and made him difficult to endure.

Eaton wrote to Torrey:

What is the matter with Rafinesque? I have defended him in New England, until I am ashamed to mention his name. His name is absolutely becoming a substitute for egotism. Even the ladies here often

. . . talk of the Science of Rafinesque; meaning the most fulsome and disgusting manner of speaking in one's own praise.

Torrey put up with Rafinesque because he respected the man's obvious intelligence. Rafinesque, born in Greece, had come to America from Sicily when he was twenty, and had begun the study of plants in a wonderful land where everything he saw was new and demanding of his enthusiastic attention. Traveling in the East and collecting plants until he had amassed a considerable herbarium, he then went home to Sicily, where he wrote up descriptions of the plants he had found, many of them new. Although he was a sound botanist in much that he did, he occasionally stooped to some fakery in proclaiming new species which were not so, or in making several species out of the minor variations in one. It was he, however, who discovered and named the prickly pear cactus in Kentucky—the main species of cactus east of the Mississippi— which today bears the name of *Opuntia Rafinesquii*.

Returning to America, he traveled into the wilderness to collect. He visited Audubon in Kentucky, and then rather astonishingly obtained a teaching position at the young Transylvania University in Lexington. He evidently tried to persuade Torrey to come out for a similar post, but Torrey was an Easterner, and he had no wish to live in the wilds of Kentucky. He was making New York the center of botanical study in the United States.

As the key figure in American botany during the first part of the nineteenth century, Torrey was a link with all the rest. When the Yellowstone Expedition in 1819 started off to the Rocky Mountains under Major Stephen Harriman Long, Torrey had been instrumental in sending Dr. William Baldwin along as botanist. When Baldwin died after the expedition had gone no farther than Missouri, Torrey sent young Dr. Edwin James to take his place. It was James who was first to climb Pikes Peak and collect alpine plants on the summit, and who brought back a large collection of new and fascinating plants of the Rocky Mountains and the Great Plains. Although Dr. James * was capable of classifying his own plants, he turned them over to Torrey. James, on his return from the Rocky Mountains, was about to set off on an expedition up the

* See Chapter 9, *Men, Birds and Adventure,* by Virginia S. Eifert.

Mississippi and did not have the time to work over the western plants. It was, Torrey wrote exultantly to another New England botanist, William Darlington, an amazing collection of more than seven hundred species, and a great many of them were new. However, since Nuttall had also just returned from the West with his own collections, it was ethical, if irksome, to wait until he published his list with its new species before Torrey published James's list from Colorado.

It was in the 1820's that John Torrey had communication with a young and inexperienced botanist in upper New York State who had written:

"I am young, without any engagements to confine me in this section of the country, and prefer the study of Botany to anything else."

The letter was signed *Asa Gray.* He was twenty-one, a medical student. He begged the privilege of asking Dr. Torrey to do him the favor of looking over some herbarium specimens and assisting in their identification. Although Torrey was then a famous botan-

Engelmann's Cactus

ist, and Gray was quite unknown, the two hit it off immediately. They wrote extensively to each other, exchanged plants and, when Gray finally started teaching botany in 1832, Torrey hired him to collect plants for him in the New Jersey Pine Barrens during the former's summer visit to Europe.

While Gray was happily collecting in New Jersey, another young botanist was doing some quiet but important work in the Midwest. Dr. John Engelmann who, in 1833, had come from Germany to practice medicine in St. Louis had, since boyhood, been fascinated with plants. At St. Louis he met Thomas Nuttall, and, fired with that gentleman's own enthusiasm, he sought to find out what grew around St. Louis, out in the Ozarks, and across the river in Illinois. Then he wandered westward, usually alone and on horseback, rambling the back trails or making a trail if there was none.

Across the Mississippi in Mascoutah, Illinois, he found another physician-botanist who was an even greater wanderer than Engelmann. Adolph Wislizenus had gone in 1839 with some traders on the Oregon Trail to Fort Laramie; to the Black Hills, back by way of the Rockies, and down the Platte to the Missouri, then to Illinois again. He and Engelmann, examining the new plant specimens, yearned for all the still unknown plants which undoubtedly were waiting to be found. Botanists like Meriwether Lewis and Edwin James, Thomas Nuttall and John Townsend with Wyeth to the Pacific, Prince Maximilian of Wied up the Yellowstone, and many others, for years had been going in that alluring western direction. It was a big country, and it still, Wislizenus and Engelmann were convinced, had not, botanically speaking, been well explored. Wislizenus, in fact, felt that the bulk of western plants, especially in the southwestern mountains and the deserts, had not even been discovered.

He could not at that time convince Engelmann that he would go on an expedition to the Southwest—although Engelmann eventually did go, and he made some splendid finds among the cacti and conifers. Engelmann, in fact, discovered more cacti and related plants than perhaps any other botanist.

Dr. Wislizenus could not stay very long in the Midwest. He would not wait for Engelmann, but went alone to Mexico to collect plants. This was in the late 1840's when the War with Mexico was in full tilt. In the mountains, the botanizing doctor thereupon was set upon by a mob of infuriated Mexicans who thought he was spying, and was imprisoned with some other Americans. Since they were not kept in a jail, but were allowed the freedom of the village, Dr. Wislizenus, making the most of the botanical opportunity, collected specimens of every plant in the vicinity before he was inconsiderately freed by the coming of Colonel Doniphan.

Then he joined the Missouri Volunteers under Doniphan and, as an army surgeon, became too busy during the conflict to have time to botanize. He was, however, one of the very first plant collectors in Mexico, Texas, and the broad valley of the Rio Grande.

These men, plus the expeditions of the boundary surveys, the railroad surveys, and the military reconnaissances, all contrived to bring to light innumerable species of American trees and flowers. On a military expedition in the Rocky Mountains and the Sierras, Colonel John Frémont, an incandescent figure both of heroism and infamy, was making history by discovering new mountains and trails in the West. At the same time, he was bringing back many new plants, including the interesting Frémontia. With much difficulty and the eventual loss of life among his men, Frémont made four trips into that harsh land of mountain and desert lying between Utah and California. The final, tragic expedition was his personal undoing. Yet, through tragedy and scandal, Frémont seemed to hold the search for plants to be an almost Holy Grail whose dedication carried him through all his difficulties.

Meanwhile, in mid-century, collectors were also going out into the wilds of Florida which remained almost as unexplored, after all the years since the Bartrams had gone there, as the most pristine wilderness of the West. The Bartrams and a few others had seen very little, actually, of the State of Florida. There were the endless, hot, palmetto flats, the coquina ledges, the swamps of the Okeechobee, the Kissimmee prairies, the cypresses and the tall, tawny grasses of the Everglades, the offshore islands and the keys, and the mangrove-tangled shores. It was a strange land of great difficulty and frequent discomfort for the traveler, yet it was full of plants which were very different, in a large measure, and more semitropical, than any to be found on the body of the continent.

John L. Blodgett, who was there in 1845, commented that in Florida a botanist must know how to wade, swim, crawl, be exposed to heat from 120 to 140 degrees, and suffer from the mosquitoes. Another collector, Edward F. Leitner, who had been sent out by John Torrey, discovered the tree which was later named

Leitneria or corkwood, but, in his botanical pursuits, he had the misfortune to be killed and scalped by the Seminoles. It was a dangerous time for botanists. Douglas had lost his life in Hawaii, Drummond in Cuba, Leitner in Florida, and Harvey Croom, who discovered a new conifer resembling a yew—the rare Torreya which is found only in a limited area along the Appalachee River in Florida—was ultimately lost at sea. He, too, had been on another plant-hunting expedition.

But this did not stop others from exploring where men might have perished. The Wilkes expedition in the South Seas and the Northwest Coast sent back huge collections. It was Asa Gray's job to work them up and publish the results. There were the Emory expedition and the Army of the West, a military reconnaissance of the western rivers; on this, a young artist collected plants and drew their pictures. Collections of plants were made in Wisconsin and the upper Mississippi Valley by C. C. Parry on the David Dale Owen geological survey. Parry later went to the Colorado Rockies and the Sierras for further exciting exploration and discoveries which included the Parry primrose, the Parry pine, and the Parry penstemon.

He and the other collectors, sending their specimens east, kept the stay-at-home botanists happily occupied. How they loved it—those plump, battered parcels coming in, travel-worn after long journeys from the wilderness, to be unwrapped with care and delight and a certain reverence, too, in discovering what was familiar, what was puzzling, and what was new. That was always the lure and the fascination, almost the vice of botany—*what was new*.

In all these expeditions and in the tremendous accession of new plants, John Torrey and Asa Gray perhaps were personally among the least-traveled of American botanists, yet were most in touch with what went on and what was being found, and were most instrumental in giving names to the discoveries. Torrey and Gray, however, finally made a trip to the West to see the mountain peaks which had been named for them in the Colorado Rockies, and to look for living plants which they had heretofore seen only as dried specimens. Torrey and his wife also made the long journey to Florida on a special mission to see the tree which Harvey

Croom had found and named *Torreya taxifolia*. With the rare Torreya in Florida, and the rare Torrey pine in California, the aging botanist was nicely represented from coast to coast.

When John Torrey died in 1873 at the age of seventy-seven, after a lifetime of botany, his death marked the end of an era in American science and exploration. He and those others had been the pioneers, the searchers, the pathfinders of botanical science. Following them would come men of an era of plant knowledge who would delve far deeper into the complexities of plant life than that first basic effort which was simply to find and classify. This had to come first; but what came later, and that which has reached no end, nor perhaps ever will, explores realms of the plant kingdom which infinitely broaden both the knowledge and the understanding of the world.

Torreya

Organ Pipe Cactus

11. A Route for Rails

Without a doubt, a man of youth, enthusiasm, and a sense of humor was a vital asset in managing the often grim and discouraging business of exploration, discovery, and the whims of politics. Amiel Weeks Whipple, lieutenant in the Corps of Topographical Engineers, United States Army, had his share of all three qualities. They not only carried him through some rather desperate as well as some annoying experiences, but also helped to prevent some other situations which might have taken place if the expedition had been led by a less thoughtful and lighthearted man.

In 1853, Whipple was appointed by Secretary of War Jefferson Davis to survey a possible route for a railroad from the Mississippi River to the Pacific Coast, by way of New Mexico, Arizona, and California.

Americans had come to the realization that this enormous country could not tie itself together as a nation until there was some efficient means of traveling from coast to coast. Until then, the only way to go from New York to San Francisco was to travel by covered wagon at great hardship, or go by ship at a great expendi-

ture of time and money. A cross-continental railroad was imperative. Also, an urgent political situation during the controversial
1850's had risen when expansion of the country produced an un-
resolved argument over slave states and free states.

For it was tacitly understood that if Southerners moved into
territories west of the Mississippi River, they would set up slave
states, while Northerners would naturally create free states. When
the railroad proposition was first made public, the territorial fight
between North and South boiled to the surface. For, if the South
laid a railroad across the southern portion of the continent to reach
California, it was obvious what these interests were about to accomplish at the same time, not forgetting the greater representa-
tion in Congress if the nation's new states expanded toward slavery
rather than away from it. Consequently when Jefferson Davis in
Washington sent out men to survey a possible route for rails, he
had to delegate more than one group. The Southern interests
would favor the thirty-fifth parallel through Texas, New Mexico,
and Arizona, while the New Englanders preferred a northern
route, possibly following much of the trail taken by Lewis and
Clark through the mountains of Montana and Idaho. The St. Louis
interests naturally favored a south-central railroad, while the
growing power of Chicago wanted one going straight west.

Since it was still extremely dangerous for lone scientists or
small, unprotected parties to go without an escort of soldiers,
many naturalists clamored for places on the railroad surveys.
There were more applicants than there were openings or funds,
for there were indeed funds to finance a limited few—it was agreed
that so large a concentration of men and effort in an unexplored
part of the country must have a full complement of scientists and
surveyors. It was through the efforts and sacrifices of the enthusiastic and dedicated naturalists who were chosen, that perhaps
more was learned about the natural resources and wonders of the
entire West during the period of the surveys than had taken place
in any previous day or which would take place in any subsequent
time.

The only way in which to find the best and most economical
route for a railroad was to send out half a dozen expeditions under

officers of the Topographical Engineers and survey a number of possibilities. The reconnaissances were to range along the thirty-second parallel from Texas to San Diego, various sections to be under the direction of Lieutenant John G. Parke, Captain John Pope, and Colonel W. H. Emory. Another route lying between the thirty-eighth and thirty-ninth parallels was surveyed by a group led by Captain J. W. Gunnison and Lieutenant E. G. Beckwith. A route from Puget Sound through the mountains of Colorado to the Mississippi was surveyed by a party led by F. W. Lander, while Clarence King directed the survey of the fortieth parallel. Another route would be examined from Puget Sound to San Diego. Still another followed the thirty-fifth parallel. This latter was Lieutenant Whipple's responsibility.

He was to have the able and reliable Second Lieutenant Joseph Christmas Ives as his assistant. Young Ives, who was soon afterward to explore far up the mysterious Grand Canyon of the Colorado, was an excellent person for an undertaking of this sort. When he and Whipple set about to choose a scientific and working staff on a budget which, including supplies and transportation, was not to exceed forty thousand dollars, they needed to be cautious. Dr. J. W. Bigelow was finally chosen as surgeon and botanist; Jules Marcou, Frenchman from Pennsylvania, as geologist and mining engineer; Dr. C. B. R. Kennerly of Virginia as naturalist and physician; and M. B. Möllhausen, a German, as a capable topographer and artist. Hugh Campbell was selected to go as astronomer. William White was an able meteorological observer and surveyor. Various secretaries and assistants, some of them political appointees with very little to recommend them, were also included.

The expedition was to report on the numbers and kinds of trees along the way which might be used as fuel for the locomotives and ties for the rails. Men were to study water supplies and the types of water to be had for the boilers as well as for the use of the passengers; surveyors were to report on the steepness of the grades, and in all cases were to seek out the easiest grade whenever there was any choice. They must, besides, note the hostility of the Indians, and do nothing, if they could possibly help it, to

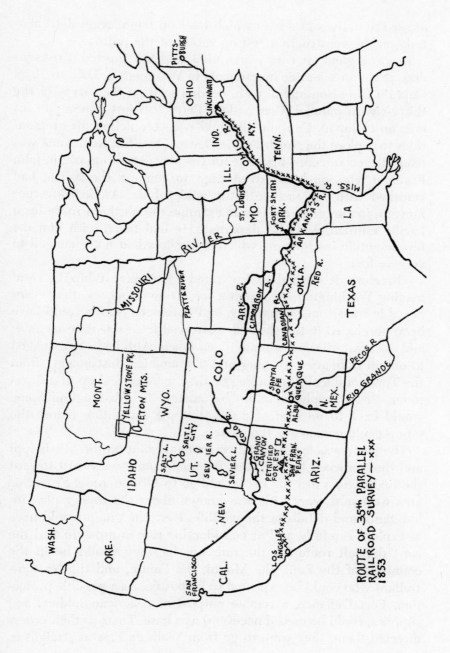

ROUTE OF 35th PARALLEL
RAIL ROAD SURVEY — xxx
1853

anger the natives in a way which later on might react detrimentally on the construction and operation of the railroad.

It was a great year for exploration, and this exerted a tremendous drain on scientific instruments in Washington, D.C. In 1853, Captain Gunnison had already headed west on the survey of the thirty-eighth parallel. Commodore Perry had just gone upon his mission to Japan. Captain Ringgold was completing his preparations to explore the North Pacific Ocean. Dr. Elias Kent Kane was ready to recommence his hunt for the lost expedition of Sir John Franklin who, on his third attempt to conquer the North, had vanished with 140 men into the wilds of the Arctic. Governor Stevens, in charge of a party to examine the northern route for a Pacific railroad, had also departed. He had taken with him the few scientific instruments which the others had not managed to acquire first.

Therefore, it seemed to frustrated Lieutenant Whipple, rummaging Washington for supplies for his own journey, that every portable transit, magnetometer, and barometer which could have been purchased in the shops or borrowed from scientific societies and observatories had already been taken. Although he managed to obtain ordinary surveying and astronomical instruments from the Topographical Bureau, where they were turned up in a storeroom, there simply were no barometers in Washington. Some would have to be made for the Whipple expedition before the group could commence the survey.

The party was to meet at a convenient point on the Mississippi and then proceed by the most favorable route westward toward the Red River, cross the Pecos, and go to Albuquerque. From this area westward, very little was known about the topography, inhabitants, and details of the wildlife. Here the group would have to explore carefully and at considerable risk in order to find the least difficult route for the railroad, for they would be in the countries of the Zuñi, the Moqui, the Paiute, and the Navajo—Indians who could be expected to be hostile. As a possible protection, Fort Defiance, a remote outpost of American soldiers and supplies, could be used if necessary as a base. Then, as their orders directed them, they were to go from Walker's Pass as straight as possible to the Pacific Ocean, probably to San Diego, San Pedro, or to Los Angeles. All three were small Spanish towns of no great importance except that they were on the coast.

While the barometers were being built, other preparations proceeded with the usual tediousness of such government-ordered activities. The Quartermaster-General's department took care of building the required wagons, tents, etc., and delivered them to the railroad to be shipped to Cincinnati, from which point they were to proceed by steamboat to the rendezvous on the Arkansas River. Meanwhile, so that no more time should be lost, Lieutenant Ives set out alone for New Mexico, by way of San Antonio and El Paso. At the latter town he was to pick up some astronomical, meteorological, and magnetic instruments which, someone had tardily recalled, had been deposited there by Bartlett and the Mexican Boundary Commission. They had been placed by the Secretary of the Interior at the disposal of Whipple and Ives.

Not until May 29, 1853, did the party, bound for the Mississippi River and the wilderness of the West, at last leave Washington. They all traveled hastily to Cincinnati to get the supplies which Lieutenant Stanley had purchased for the expedition, and added medical supplies and ammunition, together with a last-minute assortment of presents for the Indians, purchased with a small grant obtained from the Indian Bureau.

From the very beginning, Whipple's genial nature was tried by tedious delays and multiple aggravations. The supplies, including the wagons, which had supposedly been sent weeks earlier by railroad to Cincinnati, had as yet not arrived and no one knew where they were. The men did not dare delay their own departure, for the Ohio River was falling and in that June drouth there soon might not be enough water for any boats to navigate there. The expedition, moreover, depended upon steamboat passage to reach the mouth of the Arkansas River and travel up the Arkansas to Fort Smith. Bound for Memphis, they went on at once, leaving orders for the equipment to be sent as soon as possible to catch up with them—they hoped—before they started into the trackless interior.

Although the survey had originally been intended to start work at Napoleon, Arkansas, on the Mississippi, they had learned that Arkansas itself was planning to build a railroad as far as Little Rock, hence they would not need to survey that area. In the face of all the delays the expedition had suffered, Whipple was grateful for the extra time which this fact saved them.

Fort Smith, July 2. The military escort came with horses and mules, but the steamboat bearing the wagons and tents had not arrived. Marking time, they made a study of the Fort Smith area. *July 11.* A letter reached Whipple informing him that the boat carrying the vital supplies had started out, but was now stuck fast on a sandbar in the Ohio River. The vessel would be on its way as soon as it could be unstuck, which, with the low stage of the river, might take some time. This was a moment when Lieutenant Whipple had need of all his sense of humor and his patience.

He simply dared not wait around indefinitely for the supplies. He borrowed covered army wagons, tents, and food from Captain Montgomery at Fort Smith, and then set off with his expedition, moving at a snail's pace which, he trusted, would be slow enough so that the green members of the party might catch on to what they were supposed to do. He had been taken aback at finding out rather belatedly that some of the so-called surveyors' assistants, who had been sent as political appointees, not only had to be taught the rudiments of surveying on the spot, but had to be shown how to carry a chain and to use a transit before they could be very useful to anyone. He hoped the supplies would catch up by the time the men knew how to do their work.

July 14. The survey—botanically, zoologically, geologically— officially commenced. The way led through thick bottomland timber which had to be cut before the mules and wagons could drag their way through. On the second day, when the group had become rather widely separated on their various surveying operations, a heavy storm boiled over the hills. Whipple, at some distance from the rest, with several companions rode hard in the ambulance-wagon to reach the camp before the full force of the storm broke. But the borrowed ambulance was rickety, and when the wagon tongue snapped, the conveyance stopped with a jolt, sinking on its side into the thick mud, and Whipple and his officers tumbled out. On foot they hastened through the growing blackness, now shattered with lightning and made deafening with crashes of reverberating thunder, while broken twigs and leaves blew like witches on the wild and windy night.

They reached the place where the camp had been that morning. The place was right, but the camp obviously was no longer there. Tracks in the mud showed that wagons, mules, and men had moved ahead, so Whipple and the others, soaked, muddy, and furious, pushed on, now and again having to wait for a flash of lightning to momentarily illuminate the way so that they might see where they were going through the storm-beset forest. Often wading up to their hips in mud and water, they finally found the camp located on higher ground adjacent to an old plantation house. Here all was in confusion. Few tents were up, and a good many of the men, having retreated to the plantation house where it was more comfortable and dry, were having a party. The only good thing to recommend the whole trying day was the hot dinner indoors which awaited the weary Whipple and his companions.

July 26. On this day the delinquent supplies caught up; the borrowed equipment was sent back; now the expedition was in earnest. There were some difficulties along the way with the recalcitrant Choctaws, whose country they were traversing; not serious trouble, just time-wasting and annoying. Whipple by now found himself often beset with the fatalistic thought of never reaching California. But the group plodded on, slowly, because it was necessary for the surveyors to work out the route; slowly, so that Dr. Bigelow and Dr. Kennerly could collect plants and animals; slowly, so that Marcou could gather fossils.

The new wagons were constantly breaking down and needing repair. The cattle and sheep which were being driven along to provide food sometimes strayed away; the muleteers frequently got into arguments with each other. At times the mules stampeded for no reason anyone could determine, although it was suspected that the Choctaws were following along, unseen, trying to run off the cattle and mules for their own purposes. Several Choctaws had been employed as guides, but when the party finally reached the western edge of their territory, beyond which lay the realm of what the Choctaws called the uncivilized Indians, the guides refused to go any farther. One said he was ill. The other made no pretense about the fact that he was afraid of the wild people beyond, and besides, the Canadian River was so low, he said, there was sure to be a great water shortage westward, and he did not wish to die of thirst. He said gloomily: "Maybe you find no water; maybe you all die." The more he thought of his own warning, the less inclined he was to be of any assistance in that unknown land. Both Indians departed quietly in the night.

Then two of the white assistants decided they didn't feel equal to going any farther. They were delegated to take the zoological and botanical collections, made thus far, to the Smithsonian Institution, and were sent back forthwith.

One night something unseen disturbed the animals. Between dusk and dawn fifty mules had strayed away, and it took all the next day to round them up, yet three were still missing. The beef cattle broke out of their corral. If it wasn't one thing, it was something else. Whipple's patience was strained to the utmost and his sense of humor seemed to have forsaken him.

On another day they came upon two Kichai Indians who were setting fire to the tall, dry, turkey-foot grass, an Indian custom to aid in hunting. When the wind changed, the fire suddenly came whirling and snapping in red fury and choking smoke straight toward the night's encampment. In the flame-reddened darkness, the camp seemed very small and helpless against the engulfing conflagration. The men rushed to burn a fire-break around the camp, while the huge waves of flame, roaring like the ocean in a storm, came rolling through and consuming the tall grass. The hot air was full of sparks and soot; the wind was insufferable to breathe.

Wrapped in their blankets the men that night lay with their noses turned toward the ground where they could breathe a little more easily. By morning the fire had burned itself out, but travel next day was almost as bad as the fire itself. They plodded across miles of charred vegetation and the blackened earth of the big burn. It was a grim, death-filled prairie which was ebony-colored to the horizon, and still lay smoking here and there. By late afternoon the men thankfully reached the end of the Stygian landscape. They stepped into an Elysium of green grass and found pools of fresh water where the weary men and animals drew to a halt. They drank. The men threw themselves upon the cool grass, splashed water on hands and faces. The mules, unharnessed, rolled, too, and drank deeply. It was heaven.

Proceeding . . . proceeding . . . proceeding. The summer heat was severe, and in this plains country there now were few trees to make shade. The heat seemed interminable, when one evening the air turned suddenly cold and stormy as a wind boiled out of the north. The tents flapped wildly and some of the less carefully pegged ones took off into the gusty night, the men turning out hastily in their nightclothes in an attempt to retrieve their property. In the wind and rain and subsequent sleet of the norther, they pounded down the pegs and tried to get the tents up again, but the wind was so roistering and tempestuous that by morning only a few were still standing. The remainder were down, the men beneath them, trying to get warm under clammy, drenched canvas. With puddles of water in their beds, blue feet sticking out, the men felt half frozen, while the mules looked actually shriveled and purplish with the sudden cold.

When Whipple, who had managed to keep his own tent over him, simply stood there and laughed at the sight, the men were not so much amused. Everything was in a mess. Bigelow was up at dawn to sort out and dry many of his plants which had been soaked in the storm, and to put them between fresh papers. It was a relief for everyone to turn out and get moving—anything to be active and warm again—but the wind punished them all that cold day as they plodded miserably across a puddly prairie.

September. The Antelope Hills. Prairies, ravines, wagons upsetting, the best barometer broken when a wagon upended in a dry gulch. *September.* Bigelow and Whipple going out on excursions into the hills, looking for specimens of rocks and flowers, and meeting some of the largest and most pugnacious rattlesnakes either of them had ever encountered. The dim outline of the Wichita Mountains stood on the southern horizon; the Canadian River curved away to the north.

Texas . . . the Llano Estacado, almost devoid of trees, a great sea of drying grass. The botanist was the first to ascend the escarped slope to the vast mesa of the Staked Plains. He was alone in the wind and the sun, in a different world, in the huge, sky-dominated, high world of the great West. The rocks and earth were disintegrated and worn into well-defined terraces, reaching from the prairie valley to the summit. Small cedars and pinyon pines formed a miniature forest in dark patches on the landscape, and there were so many new plants that, when he finally re-

Bigelow's Aster

turned, his collecting case was literally crammed with plants and blossoms, many of them of unknown species. The geologist Marcou was down on his knees with his hammer and bag, collecting both the Cretaceous rocks and the pieces of limestone which contained fossil shells.

As the sun set clear and calm, the atmosphere was ineffably crystalline. The storms were past, the trials were over, the evening benign. It was as if the whole expedition now took a second breath and straightened its collective spines. The men realized that now they were on the brink of new discoveries and new lands, as well as undoubted new dangers. But surely, all would go well in this clear, mountain-flavored atmosphere.

Next morning, they met parties of Mexicans and Indians riding north and heading for the Comanche country, all of them evidently afraid of something they were fleeing in the south. They paused long enough to explain that there was trouble back at La Mesilla. General Trias had come with troops, and wild rumors were about that Santa Anna himself was coming with five thousand men.

The American soldiers were wonderfully revived. Traveling along a difficult and often monotonous trail, fighting fires, making interminable surveys, shooting rattlesnakes, losing tents in a storm, and collecting plants and rocks were all very well, they supposed, but the prospect of some trouble with the Mexicans and Indians sounded much better. So Santa Anna was back, was he? they were saying happily. You couldn't keep that old devil salted away forever. He might have been bested by the Americans when Mexico City fell in the recent war; he might have been sent off to exile; but a man like that—no, you couldn't keep that one down. He'd be back, and if now he had five thousand men at his call, then the old general hadn't been beaten yet, nor would he be until he was laid cold in his grave, and his wooden leg with him.

Whipple scouted the idea that the Mexican general was out of exile or in the vicinity, or that he had a huge army ready to invade and reclaim New Mexico and Texas, which had so recently belonged to Mexico. But he could not quell the high spirits. The chance of a collision with Mexico lent a certain spice to the long days.

The night's camp lay near a prairie dog town which served to take the men's minds away somewhat from the talk about Santa Anna. Some of them, backed by the naturalists, decided they would try to dig out a prairie dog and discover how deep its burrows ran. This activity, however, turned out to be discouragingly hard work. The soldiers dug for some time and then gave it up. Next they used long, pliant willow poles to probe far down inside. The spiral burrow, they found, went down about five feet, but nothing had thus far routed out any of the inhabitants.

Undaunted, the men would not give up. This was a challenge. They tried the water treatment next. They sluiced in six pails of water brought with some effort from a distant stream, but it sank into what seemed to be a bottomless burrow. In two minutes all trace of the liquid had vanished. Said Whipple in amusement: "Neither the rattle of a snake, the hoot of an owl, nor the chirrup of a dog, gave token of life within. Nothing but a swarm of crickets seemed to have been disturbed." He wished that his men showed as much enthusiasm for their duties as they did for digging out a useless prairie dog.

That night there was a great commotion in the prairie dog colony. The little animals were all chattering and chirruping until the men thought that the coyotes must be attacking. When morning came, however, the burrow which had been rudely dug into and flooded had been neatly stopped up and sealed over with fresh earth. Only a few prairie dogs sitting watchfully on the more distant burrow mounds barked now and again in disagreeable, accusing tones.

November 7. Albuquerque, New Mexico. Lieutenant Joseph Christmas Ives had reached here a month before, and it was now a considerable relief to both Ives and Whipple to make contact at last, since neither had had any message from the other during their separate journeys. Ives had spent his time in securing supplies and interviewing guides for the coming ordeal in the untraveled wasteland and desert mountains. The area was but little known. Only the Sitgreaves expedition two years before, and part of the Boundary Survey, had come this way. It was, therefore, imperative to have a guide or two along who had been over the territory, or who had at least some experience in the terrain and with the Indians. Whipple, who had heard of the excellent French guide Antoine Leroux, who had accompanied Sitgreaves in 1851, wished this man might now be available, but this was palpably impossible since Leroux, Whipple knew, had been commissioned by Captain J. W. Gunnison as a guide on the Central Pacific Railroad survey. This expedition was even now pushing its way through the valley and desert of Utah. In the case of the invaluable guide Leroux (as with the scientific instruments), Whipple was feeling as he approached Albuquerque, that he had been the last man to be considered or to have any right of priority. All the best had been made off with before his needs were known.

What Whipple, before meeting Ives, did not know was that the invaluable guide had in fact just come back to Albuquerque. He was there now; Ives had already hired him for the work at hand.

But the terrible fact which Whipple, Ives, and Leroux did not know at that time—and perhaps it was just as well that they did not—was that Gunnison was dead. His botanist was dead. So were his topographer, his two guides, and three of his soldiers. The Paiutes had killed them. And Whipple and his men were now about to start into the country of the Paiutes.

It had happened on October 26, more than a month after Leroux had gone back to New Mexico. Gunnison's survey had almost finished its work for the season. His orders had been to take up winter quarters in Salt Lake City under the protection of Brigham Young and the Mormons, and start out again in the springtime for a further survey of the proposed railroad line from Salt Lake City to the Pacific Coast. In October, Gunnison was along the Sevier River, north of dry Sevier Lake in the desert of central Utah.

Now and again there had been some mild trouble with the peppery Paiutes, who were a touchy and often irritable tribe. However, nothing very alarming had taken place. It had been the custom of the party to separate in several groups to make a wider survey, and to feel so little apprehension of Indian danger that when Gunnison took a small party with him to explore the river, no one was worried. He had with him R. C. Kern, one of the exploring Kern brothers, who was an expert topographer and artist; F. Creutzfeldt, a German botanist; two guides, and four soldiers. On October 25 they had gone up the river and were expected back next day.

But when the twenty-sixth passed without their return, Lieutenant Beckwith, second in command, grew uneasy. Not until the following day did a lone, exhausted man, half dead from his ordeal, stagger into camp with the news that the Paiutes had attacked Gunnison's encampment and all, he thought, had been killed, all but himself. He would have died too if he had not lain wounded in the bushes all night. He had been half dead with pain.

When he had finally roused himself and raised his head cautiously out of the creosote bushes, he had seen that the camp was empty. Getting to his feet, he discovered trails of blood which showed where bodies evidently had been dragged out to the desert. Beyond, the blue-shadowed House Range stood up in pink and amber serenity in the peaceful light of the autumn morning. He heard wolves howling as he set off to regain the main camp.

Beckwith and his men, with great foreboding, hurried to the site of Gunnison's camp. There the Indians had made off with everything and there was no trace of the bodies. It was so nearly dark when the rescue party reached the site that they dared not pursue a further search until the morning. So Beckwith and his men lay fearfully in that fateful camping place, afraid of what the darkness might hold, fearful of what they would find in the morning. It was an endless-seeming night, yet no Indians came, and few sounds were to be heard except a small owl crying plaintively somewhere in the starlight, and the wolves squabbling over something at a distance. The thought of what they were no doubt fighting over chilled the men's vitals with a sick despair.

What was left when the Indians and the wolves had finished was found in the pungent chamisso and sage next morning. The bodies had not been scalped—that was not the way of the Paiutes. Instead, the arms had been hacked off at the elbow, and the bellies had been cut open and disemboweled. Then the animals had come. What remained was not a sight for any man to see, and Beckwith and his men were so shocked and sickened they could scarcely take over the task of digging graves for the remains of their leader and his companions.

Brigham Young in Salt Lake City was horrified. He called the Paiute chiefs to account, and they professed to be shocked also. They said it must have been the work of irresponsible boys who had done this thing in retaliation for people of their own who had been killed by white men. Not by these white men, certainly, but to the Paiutes that didn't matter. White men were all one to them. With some difficulty and strong persuasion, and only after rewards were offered, Gunnison's notebooks and records, Creutzfeldt's plant specimens, and Kern's maps and sketches were returned. The scientific instruments, broken and quite ruined, had been recovered from the desert. The puzzled Indians, not knowing what to make of these strange objects, had tossed them away, but first had taken care to break them so their evil spirits would depart.

Creutzfeldt had found new plant species in Utah. Several would be named after Gunnison, as would the awe-inspiring Gunnison River and its deep canyon in Colorado. The Kern River in California, a galloping and beautiful mountain stream in the Sierras, would commemorate the three adventurous and ill-fated Kern brothers.

Such memorials were well enough, but they did not bring back the men who had died in the name of science. After 1853, no sci-

Gunnison's Gilia

entific expedition went into the West without the knowledge of what had happened to Gunnison and to Kern and to the quiet, kindly botanist Creutzfeldt, who had come from his own land to study the plants in the marvelous and murderous American West.

But word was slow in getting about to the few places occupied by white men in these remote mountains and deserts. Therefore, in November when Lieutenant Whipple was delighted to learn that the capable guide Leroux had come to his service, his joy was unmarred by knowledge of the recent tragedy in Utah. At last, he may have felt, things might be turning his way, while Leroux, ready for another job, welcomed the chance to go along with Whipple and the reconnaissance of the thirty-fifth parallel.

The expedition left Albuquerque and headed into the Indian country. The pueblos on their mesas, fascinating in their remoteness and their other-worldliness, were on the itinerary to be examined and reported on. When, late in the afternoon, they neared the ancient pueblo of Laguna in the red sandstone country, the late light, shining on the remote pueblo, glowed through a pink-gold haze which gave the place a dreamlike quality. Germans in the party nostalgically spoke of how it reminded them of a terraced village on the Rhine.

Visiting the pueblos was a strange glimpse into ways of life of people unchanged for several thousand years—all but the Zuñi, where the smallpox which the white man had brought to the Indians was just then ravaging the high, mesa village with a modern pestilence. The soldiers, although they had all been vaccinated, were worried. They did not stay long at Zuñi, where so many of the people had already died and where others lay ill in every house.

The expedition had no sooner gone away from Zuñi than an incident stirred the camp. One warm November day, some of the teamsters craved a drink. They knew very well that a keg of spirits reposed in one of the supply wagons, and so they secretly helped themselves to the contents. Several of the Mexican servants were cordially invited to join the party but, wide-eyed, they declined and in fact turned quite white. They knew what was in that keg—it was alcohol for preserving the zoological specimens, and they remembered clearly that when the party was still at the Choctaw agency, Dr. Kennerly had ordered that the liquor, for further preservative powers, should be liberally spiked with arsenic. By the time the frightened Mexicans could stammer out the reason why they didn't care to have any, the teamsters had already swigged down a good deal of the stuff, and now they immediately began to feel desperately ill. Moment by moment they grew sicker and more terrified. Still undecided as to whether to call Dr. Kennerly and thus reveal how they had sinned, or agonize in silence and suffer a horrible death, they staggered into the camp. Now so ill that they could no longer care about punishment, if only they might be spared, they tottered to Kennerly's tent. Dr. Bigelow hurried in after them when he heard the groans and commotion. Green and nauseated, scarcely able to hold up their heads, the culprits gasped their confession.

Dr. Kennerly and Dr. Bigelow, much relieved, broke into

laughter. Whipple, who had followed to see what had happened, joined them, while the sufferers, feeling no better at all, could not understand such inhuman lack of concern for their plight. No, the doctors gasped between gusts of mirth, the alcohol itself was the only poison they had taken, but because it had been dosed, not with arsenic, but with ipecacuanha, a powerful emetic, they would very soon be relieved of the whole thing and of their misery at the same time.

There was a good deal of joking in the camp after that as the anguished teamsters, rushing out behind the tents, proceeded to vomit until they were rid of everything they had taken. The incident had its good points. Whipple and the naturalists knew that not a man, after this, would broach a keg of spirits. There had been occasions on other expeditions, both on land and at sea, when the specimens preserved in alcohol had been quite ruined when the group reached home at last and the kegs were opened. Someone, it was frequently found out too late, had drunk off all the alcohol, thus leaving the specimens of fish, fowl, or beast to deteriorate in the bottom of an empty container.

Moving slowly through the desert, they found it a strange realm of cacti which were a discomfort for men and animals but an excitement and a pleasure to the botanist. When the party moved up into the hills and to the land of tall pines, they found plentiful game to augment the tiresome army rations. Water was abundant and good. Up in the hills Bigelow collected new specimens of mistletoe and lichens, and found a large tree Opuntia with great flat pads of leaves and yellow fruits which had never been described before. For that matter, most of the plants he was now finding had not been seen by botanists.

November 29: Camp 73 after leaving Fort Smith. "We have now broken away from all communications with the civilized world; and, for the first time on this trip, have entered a region over which no white man is supposed to have passed."

They camped at Jacob's Well, or, called by the Indians, *Wah-nuk-ai-tin-ai-e*, from which Navajo trails radiated in all directions. The well was a pool of water some thirty yards in diameter; evidently deep and enduring, it was surrounded by a fringe of slender tules. Until now the party had seen no Navajos, a warring tribe which was not to be trusted; only their trails and fairly fresh tracks showed that they were in the vicinity. But the next day at sundown as a pillar of smoke rose from a great mesa, two superb horsemen galloped with a flourish into the camp—Navajo hunters coming back from the mysterious depths of the Canyon de Chelly. They did not stay. They had learned that the white men had but lately visited the pueblo of the Zuñi where the smallpox was very bad. They did not care to have any contact with stupid white men who had so rashly come close to the dreadful disease. The Americans did not enjoy being reminded of Zuñi and of the sick and dying people there, for several of the soldiers had already come down with varioloid, a light form of smallpox. Yet if the fear of the disease kept the Navajos away from the camp, it had that, at least to recommend it.

Westward . . . into a strange region imprinted with the pictures of the past . . . a place where great fallen trees of a prehistoric forest had all turned to stone. It was an astonishing sight, an incredible discovery. They could not understand how such large trees could have lain there for so long, trees which could have become jasper and onyx and agate marked with bands of blue and purple and red and cream. The stone and the colors had replaced the original fibers of the wood. The Whipple expedition had discovered the petrified forest of Arizona.

Jules Marcou, quite dwarfed by the fallen stone trees, stood bemused. He examined the glistening, jasper-impregnated wood which was no longer wood but stone of great hardness. He tried to conjecture the size of the original trees, wondered what sort of forest they could have created, wondered when they had lived. Then he wondered if the forest had grown here from the beginning, or if the mighty trunks, carried away in a great flood, might not have grown elsewhere and been brought here ages ago to petrify. He and the other scientific men on the expedition were awed by what they, no doubt first among white men, had come upon in that harsh, cruel yet beautiful desert. When one surveyed for a railroad, there was no way of knowing what the landscape might offer.

"Quite a forest of petrified trees was discovered today," Whipple wrote, not realizing the magnitude of the find, "prostrate and partly buried in deposits of red marl. They are converted into beautiful specimens of variegated jasper. One trunk was measured ten feet in diameter, and more than one hundred feet in length. Some of the stumps appear as if they had been charred by fire below before being converted to stone. The main portions of the trees have a dark brown color; the smaller branches are of a reddish hue. Fragments are strewn over the surface for miles. Now the soil produces no timber; the scrub cedars have disappeared. For the last three days dry twigs of chamisa have been the only fuel available for camp fires."

They named a dry river-bed Lithodendron Creek, in honor of the petrified forest—a forest which offered no fuel, only stone to men in danger of their lives.

At this point they were not many miles from the Grand Canyon of the Colorado, and they had no doubt heard of it, or at least knew rumors of its immensity and awesome depths, but since they were surveying for a railroad, and railroads had no business in canyons of this sort, they evidently did not go to investigate it. Joseph Ives, however, evidently enthralled with that place of mystery and great depths, led a later expedition by boat up from Yuma on the Colorado River, and followed part of its dangerous course. He, first among white men, scaled some of the great precipices of the canyon, some years before Major Powell traversed it by boat. Ives considered it to be of no possible use to any man.

Volcanic scoriae . . . a majestic scene at dawn as sunlight cast a glow over the snowy San Francisco peaks rising from their dark pine forests. Christmas Eve . . . Christmas Eve at Cosnino Caves up in the mountains, a pleasant place with plenty of fuel and food. . . . They celebrated the occasion with a party, complete with fireworks of utmost magnificence, for they set fire to several tall, isolated yellow pines near the camp, and with a roar the trees flared in a tremendous gush of ruddy flame against the night sky. The illumination lit the surrounding forest, the mountainside, and the camp, and then died away in glorious showers of sparks. Some Navajos who had come to pay a visit were delighted. The men sang, performed duets, gave recitations in which they had the privilege of saying what they pleased and of being pleasantly insulting to everyone.

The Mexicans performed a solemn nativity tableau on the mountainside. The evening wore on, the temperature dropped, and by dawn it stood at zero, so that the men woke to a scintillating, frost-ornamented Christmas Day. It was Christmas Day . . . and, considering his middle name, perhaps it was the birthday of Lieutenant Ives. Some of the pleasure of what had been declared a day of rest was lost when it was realized that nineteen mules had strayed away during the night and must be rounded up. The visiting Navajos, who also had gone, were suspected.

January . . . the whole of January . . . moving slowly through a vast landscape into the mountains, over deserts of appalling desolation, and with difficulty crossing rivers. *February* . . . up a rocky pass which, because of the numbers of cacti everywhere, was christened Cactus Pass. Country of the bighorn sheep . . . the sheep, with airy leapings from rock to rock and never missing, made the weary mules look like creeping toads and made the men feel sodden and earthbound.

Bigelow was enthralled with the many species of cacti in this strange, arid country. Although some of them had been discovered and identified elsewhere, following the Boundary Survey, and by Dr. John Engelmann and Dr. Adolph Wislizenus—who were now the authorities on these plants—Bigelow realized with a thrill that he himself was finding many cacti which even these indefatigable botanists had not seen. This broad, arid, sun-bathed region was now almost entirely a garden of cacti and other desert plants—palo verde, yuccas, desert willow, Joshua trees, ocotillo, and many more. The cacti themselves ranged from the gigantic Sahuaro which were like trees, often forty feet tall, in whose bodies the little elf owls nested in holes made by the Gila woodpeckers, to the small hedgehog and stout barrel cacti. Numbers of Opuntia, Cereus, and other species were both confusing and exciting. Although these plants made a thorny path for men and mules and were a trial to the men who had to survey the terrain, the way was uncommonly pleasant for Bigelow and Kennerly. So excited were they that soon Whipple, Ives, and some of the others in the party, including the Mexican muleteers, joined in the hunt. The day's labors became almost a game as each man turned in his own contributions of cacti. The botanist could only wish that the plants could all have been in bloom, not largely dormant in their winter state. Some, however, were beginning to put out buds and a rain would open them quickly in the glory of the desert spring.

The men were also the unhappy discoverers of the uncomfortable cholla cactus, which had the bad reputation of throwing spiny

Whipple's Cactus

pieces of itself on the ground all around the plants. A painful structure in the extreme, the spines broke off at a touch, and the ground was so littered that the mules were in agony; the vicious spines even thrust into the men's shoes. Vicious or not, cholla was a new species and was duly recorded. Home of the cactus wrens which built their nests in these cul-de-sacs of thorns, and haunt of the road runner and desert lizard, the cholla was part of its harsh environment.

The artist Möllhausen, as he had done all along the route when new specimens were found, was making highly detailed pictures of the cacti for, since pressed specimens of most of these were quite impossible to make, drawings were essential for later identification. Möllhausen's illustrations of the cacti were the basis for the beautiful engravings by Isaac Sprague in the report when it was finally published. Sprague, one of the best botanical artists of the day, had traveled as a young man with John James Audubon on a journey up the Missouri River to the Yellowstone in 1843. Audubon had named the Sprague's pipit in his honor, and a wild flower, the western Spraguea, was also named for the botanical artist.

The Whipple expedition passed through the country of the Mohaves, of the Navajos, of the Paiutes, and there was no trouble with the local inhabitants. Yet Whipple, day and night, was uneasy. He had insisted that the men maintain good relations with the Indians at all times, no matter how hard it might be to do so. But disaster nevertheless finally struck.

One of the Mexican muleteers one night was left behind. Torrivio, a man they all liked, and a conscientious one, had gone back over the trail to look for several mules which had strayed; he could not rest until all his charges were safe. However, Torrivio did not return. When Whipple, Ives, and some of the soldiers went back to look for him next morning, they found what was left of the unfortunate muleteer in a thicket of palo verde, and unmistakable signs of Paiutes around. And what had happened to poor Torrivio, every man knew, might also happen to him suddenly and without any warning. The men, without being ordered to do so, refrained from scattering too far from the camp or from the main body of soldiers.

On March 14, they met a party of Mormons on the Mormon Trail bound for Salt Lake City, the first white men the Americans had seen since last autumn. The officers and scientists were delighted to talk to them and get the news. But the news was not good, for it was from the Mormons that the surveying expedition finally learned about the fate of Captain Gunnison and a portion of his party. Every man in the group, recalling what had only recently happened to Torrivio, fervently wished he were out of this terrible land.

But the end of the thirty-fifth parallel survey was in sight. Beyond the alkali flats and the soda lakes in the Mohave Desert of California, the mountains of the coastal ranges rose tall, stood blue-purple and serene in the morning light, with the white of the alkali below shining innocently, like snow, in the sunshine. Desert quail whistled and ran in troops under the gray sagebrush, crested flycatchers screeched along the edge of the mountains, and the cactus wrens were singing their full flood of spring song as the expedition wearily contemplated the steeps of Cajon Pass, which lay ahead. There was no easy way around these mountains. The surveyors realized with regret that a railroad would have to be managed up this pass and the grades accomplished somehow.

In a gentle, life-giving rain next morning, they left the cruel and colorful desert and started up this pass, up into the live oaks and among the California sycamores, over the top, and down to the greener western slope. Forest trees, blossoming shrubs, and wild flowers in a suddenly different world manifested the California springtime.

They camped at Cucamonga. They went on past the Spanish rancherias, past vineyards, through all the lovely flowering of spring and its ineffable fragrance, in a soft, serene, and perfumed atmosphere which is peculiarly that of California. They moved southwestward to the sea.

. . . and then, from a slight eminence, we looked upon the valley and the city of Los Angeles . . . and entered the city . . . with vineyards, orange and olive groves, peach orchards, gardens, and cornfields. Along dismal-looking lanes were scattered piles of adobe houses and the intervening spaces were lined with mud walls and cactus hedges. But as we proceeded towards the plaza, the somber character of the place nearly disappeared before the march of American improvements . . . the buildings were formerly of one story, with a broad piazza in front. But houses are being erected in more modern style, and many white tents in the suburbs marks the spots where newcomers design to build. The population is said to be 3,000, and is rapidly increasing.

They were finished. They were now within the jurisdiction of the men of the Coast Survey. When the fog lifted next day, there began an auction of all equipment, a sale at which the travel-worn mules actually brought more than a hundred dollars each. Then, with the transaction concluded, the officers, in a driving rain, rode the stage coach over an execrable road to San Pedro. It took them eight hours, although the town was only twenty-five miles away.

The officers and naturalists, all but Mr. White and Mr. Sherburne, who were so enchanted with California that they intended to stay, thereupon took ship from San Diego to San Francisco.

This was the main port on the West Coast where they must wait for a ship to take them to Washington. Bigelow, however, stayed in California until June in order to botanize more extensively along the coast and go up into the Sierras to see the incredible Sequoia trees of which he had heard rumors. His collections finally totaled more than sixty new species and many new genera.

After all the survey reports were in and had been evaluated, it was still impossible, politically as well as geographically, to decide on a single route for a railroad. Eventually, lines were laid along all the survey trails. Whipple's route was largely occupied by the Atchison, Topeka, and the Santa Fé, as it still is today.

Another result of the surveys was a set of twelve remarkable, elegantly illustrated volumes, published by the government, comprising the detailed reports of the *Surveys for a Pacific Railroad*. Poorly concealed beneath accounts which attempt to be impersonal, the books also contain the very personal adventures of a number of hardy men who discovered some of the infinite and dramatic details of the American West.

Beckwith's Violet

12. The Teton-Yellowstone Wilderness

Grizzled old Jim Bridger had scratched himself, and with distaste he had looked up at the mountain. Nobody, he averred, at least no white man, and certainly no Indian in his right mind, had ever climbed the Grand Teton, or ever would.

The Indians knew better than to waste time at it. Besides, what was there to be gained? Nothing lived up there in those windswept snows of the sharp peak, and there was neither food nor treasure to lure a man. Why bother to go up? As for the white men who seemed possessed with the urge to climb anything that stood in their way, none had as yet managed the Grand Teton. Nor were they likely to do so, Jim Bridger believed, expressing his opinion with the finality of a man who had lived in this wild western wilderness of the Yellowstone and the adjacent Tetons, and had long since taken its measure.

Jim Bridger and the other mountain men of his time felt a personal and proprietary interest in the Tetons and the Yellowstone Valley. They themselves had found these places and had told

about them, but their stories had in the main not been at all believed. Mountain men were notorious liars, and their stories of
boiling mud pots, paint-colored rocks, enormous waterfalls, and
fountains of hot steam by which you could set your watch, were
all very obvious fabrications worthy only of laughter.

The Lewis and Clark expedition had passed very close to the
Yellowstone Valley when, on a side trip led by Clark while Lewis
was exploring the area now known as Glacier National Park, they
had followed the Yellowstone River, but they had not gone far
enough. When the expedition was on its way home down the Missouri, John Colter, one of the best hunters of the party, had asked
to be released from duty so that he could go back into that country
and explore it. It was John Colter, then, who discovered part of the
incredible Yellowstone Valley.

The Indians had stayed away from it as a haunted and accursed
place. "Land of Burning Mountains," the Shoshones called it. It
was a place of evil spirits, and although game was plentiful there,
few or none of the Indians ever went into the valley to hunt. Later,
when John Colter came back to tell of what he had seen in 1807
and 1808, people disbelieved him. They jokingly called his non-
sensical discovery "Colter's Hell."

Jim Bridger had come upon the place by accident in 1830, but
because his own stories were generally always so gaudy and highly
embroidered anyway, people usually felt that this new one was
just another Bridger elaboration. It was only reasonable that no
one would be gullible enough to believe a man who told of a
mountain with a six-hour echo. When he had camped in that area,
Bridger had solemnly related, he had yelled "Time to get up!" in
the direction of the mountain when he was ready for bed. And
then the mountain, six hours later, had obligingly thrown back his
words to waken him. His stories were too much for any honest man
to believe. Nevertheless, Bridger, like Colter, had indeed seen the
Yellowstone.

From his first knowledge of it, that country and its mysteries
had also fascinated Frederick Vandeveer Hayden. In 1856 he had
been connected with a government expedition to explore the lower
Yellowstone River, under the command of General G. K. Warren,

U.S. Corps of Engineers, with James Bridger himself as guide. But during the time in which the expedition was delayed and delayed again, General Warren was superseded by Colonel William F. Raynolds and the expedition was on its way in 1858 and 1859, with Hayden as geologist. Government exploration in the nineteenth century went ably equipped with some of the best geologists, topographers, botanists, and zoologists of the day, often several of each, or sometimes, if appropriations were low, with one man to fill several positions. Hayden, one of these excellent-all-round naturalists, was a boon to the government budget.

The first Yellowstone expedition did not manage to penetrate very far into the area. The mysteries still lay behind their mountain barriers. Not until the summer of 1870 was another government expedition, led by General Henry Washburn, sent out to discover at last what, if any, truth lay in the strange stories still circulating about the Yellowstone. Washburn himself, leaving Helena, Montana, was obviously skeptical and impatient with the assignment. He could have thought of better things to do with his time and the government's money. Colter's Hell and Bridger's Wonderland were too fanciful for men to accept. But some of Washburn's men were eager to go into that valley and find out its truth. One was Fred Hayden and the other was N. P. Langford, two young men who had seen enough of the West to realize that any tale of those hidden valleys might very likely be true.

The party rode south into an incredible country like nothing else on earth, and into some of the strangest experiences which perhaps any men had ever known in America. The rock formations, the exploding geysers, the splendid colors, the paint pools, the thundering waterfalls, the gorges, the mud pots, the boiling springs, the wildlife—the birds, the mammals, and the carpets of flowers—the high mountains with snow on their summits, walling this wilderness . . . it was all a special brand of American magic.

One night the men, all but surfeited with astounding sights, sat around a large fire which lit up the tall, surrounding pines and talked of what they had discovered. Some discussed how this wonderland might be used for profit. It was obviously not for agriculture; only a show place, the landscape was impossible for anything else but sightseeing. The first thought of several of them was to get enough money among themselves to buy the Yellowstone land from the government and exploit it.

But others objected. Judge Hedges disagreed strongly. This wonderland, he said sternly, was not the kind of place to be owned by any man or group of men. It was part of America, and it must be set aside forever for the inspiration and pleasure of the people. Although America in 1870 had no national parks and up until this moment had never seemed to have taken much interest in preserving a wilderness which had been the obsession to tame and destroy, the idea for the first national park was evidently born on that night around that campfire in the valley of the Yellowstone.

Young Langford was bubbling with the plan. While the flames died down to coals and the coyotes wailed in the distance, he and the others talked long into the night. When the men were all back in Helena, they worked toward the idea of a national park. Judge Hedges wrote newspaper articles and Langford lectured in eastern cities, wrote articles, conferred with government men in Washington.

The efforts fell flat. People in the East frankly did not believe all they read in the papers or heard on the lecture platform. There were no pictures to prove anything, only wild tales of men who obviously had something to sell, although the public could not be sure what it was. Langford, next to Jim Bridger himself, was called the best liar in the whole Northwest. Chilled, he simply waited.

Then, to settle the argument, the U.S. Geological Survey in 1871 sent a party into the field, really exploring the Yellowstone, and they had with them the best photographer of the times—or since, for that matter—William H. Jackson. He hauled his cumbersome equipment on mule or horse over mountains and up cliffs to take pictures which have surely not been improved by anyone else in the years since he took them. The photographs and the reports of the scientists on that expedition of 1871 were to prove incontrovertibly the existence of the fantastic wonders of the Yellowstone.

The expedition was the crucial one to decide the fate of that area. Leading the scientific crew was Hayden himself, with James

Stevenson as his principal assistant. In addition to Jackso[n]
photographer, an artist was sent along to supplement his wo[rk]
carry on if the camera equipment failed or unfortunately fell
a precipice. There were men seeking out information on the [min]
eralogy, the zoology, the meteorology, and the botany of the Yellowstone—Professor G. N. Allen and his assistant were the botanists.
And with this party of nineteen dedicated men there came a noted
painter from Philadelphia, Thomas Moran—they later named Mt.
Moran in the Tetons for him—who went along to paint, in full
color, the remarkable scenery of the Yellowstone and the Grand
Tetons. In addition, fifteen men went as teamsters, laborers, cooks,
and hunters.

The journey commenced at Ogden, Utah, where the mules,
horses, wagons, and food were unloaded at the railhead on June 1,
1871. It was a tedious trip before ever reaching their goal, for the
expedition must map the route, make a careful survey of mountains and passes, collect birds, plants, and mammals, look for minerals and collect Indian artifacts. That was the tediousness, as
well as the fascination, of the scientific expedition. It was not
expected to make speed, not when Professor Allen had some new
flowers to secure, or new items were to be added to Mr. Schön-
born's maps, or when Mr. Carrington was ranging out to follow
and triumphantly bring back a new bird.

From their base at Boteler's ranch, they then pack-trained into
the valley of the Yellowstone and arrived in late July. They stayed
through August; this was the perfect time for the great abundance
of flowers. The weather was glorious, the scenery, of course, absolutely incomparable. Hayden, in preparing his report, brought out
all these facts. His purpose obviously was to point them out to the
prospective tourist. "The finest of mountain water, fish in the
greatest abundance, with a good supply of game of all kinds, fully
satisfy the wants of the traveler, and render this valley one of the
most attractive places of resort for invalids or pleasure-seekers in
America." It was a comment which certainly was not the customary addition to a sober geological report, but Hayden was desperate, and he had to add all the points which might sell Yellowstone as a national park to the Congress and to the President.

It was almost impossible to describe this place so that people could see it as the men experienced it. The incredible rock formations, the massive wrenching of earth and stone which had created the valley—how could one adequately describe the Grand Canyon of the Yellowstone, or the falls? Standing on the margin of the Lower Falls, Hayden looked down into the gorge which was an immense chasm in the basalt, whose sides rose twelve hundred to fifteen hundred feet high, of the most brilliant colors, the rocks weathered to an infinite variety of forms. Only Bill Jackson's camera and the brush of Tom Moran could capture this.

Moran was in ecstasy. This was what he had come for, this incredible magnificence. He sat himself down on the rim and began to paint.

At Yellowstone Lake they were again all but speechless, and Moran once more was busy. As the expedition camped at several spots around the lake, exploring parties went out to map, others to collect plants. On July 31, Hayden took the topographer, the artist, and the mineralogist with him to seek out the geyser basin of the Fire Hole River, which had been vaguely reported but not explored. Since they had no guide, they got themselves well entangled in a network of fallen trees on the slopes. The horses stumbled incessantly and threw their loads. The men, with a good deal of hard work, bruises, scratches, and aggravation, traveled thirty-one miles, at least a dozen of them through the giant jackstraws of the fallen pines. And at the end of the day they came to the strange alkaline valley of the Madison River, a nightmarish place. The Fire Hole Basin was full of steam vents and poppings and rumblings, with colors unbelievable, and the smell of sulphur all about. The whole area smoked like a factory town on a winter's morning.

The men could not hope to report in detail on everything in this huge area of the Yellowstone, only stress the highlights, the most startling and remarkable of phenomena. Of them all, it was the geysers, perhaps, which fascinated them most.

Some were steady gushers, others intermittent spouters, and, until the men learned the rhythms governing each one, some of these suddenly shooting jets of steam and water invariably sent the party scrambling out of the way. By August 5, when they had camped among some of the greatest geysers in the world, they had still not been there long enough to take the show calmly. The preliminary warnings with mounting internal pressure, the rumblings, the sudden roar of steam and water often gushing two hundred feet into the air, were endlessly exciting. The Grand Geyser, the Bath-Tub, the Punch Bowl, the Giant, the Giantess, and Old Faithful were all named by these men on this expedition of 1871.

The journey was completed in September. As the party disbanded, most of them hurried eastward on the railroad, bound for Washington to write up their reports and present them hopefully to the government. More than ever before, Langford, Hayden, Stevenson, and the others knew they must present these reports and their arguments in such a convincing manner, augmented and illustrated by Jackson's photographs and Moran's paintings, that there would be no possible way for the Yellowstone to be lost.

They must have been persuasive enough. Early the next year, Congress passed the bill for which the men had worked so long and so desperately, and President Grant signed it on March 1, 1872. The Yellowstone Valley became the first national park in the United States. There, N. P. Langford, later called "National Park" Langford, serving as its first superintendent, worked in his beloved wilderness for five years without pay. In this area, sixty-two miles long and fifty-six miles wide, encompassing parts of Idaho, Wyoming, and Montana, Langford was where he most wanted to be.

It was after all this that, with less pressure, the men could return to the Yellowstone for a more painstaking and detailed exploration and collection of plants and animals. In 1872, when Frederick Hayden led still another expedition out there, exploration of the adjacent Teton Mountains south of the Yellowstone was included. With him went a young and dedicated botanist, John Merle Coulter, who would later go on to a lifetime career in American botany in which his son would follow him. Coulter, it might be said, cut his botanical teeth on the Yellowstone expedition.

In order to explore more widely and efficiently from the geyser country and innumerable unknown valleys and mountains of the region, to the mountains and glacial lakes of the Teton Range, the party separated into several groups. Langford and Stevenson, leading the group into the Teton area, came at last from the rough, mountainous country to the west, to the base of the tremendous, uprearing bulk of the Three Tetons which dominated the landscape above the valley and the lakes. The Grand Teton, largest of all, loomed as tall and dramatic as the Matterhorn—a rugged, blue-purple, snow-covered, glacier-furrowed, tremendous peak challenging all men.

From the camp on West Teton Creek, the men looked up at the Grand Teton. Old Jim Bridger had said it long before and always maintained it, that the Grand Teton would never be climbed by any man—by mountain goats and mountain sheep, perhaps, but not by men. No Indian would expend that much energy, and no white man, willing or not, would ever be able to scale those precipitous, wind-swept, white-mantled heights. But Stevenson and Langford needed to climb the Grand Teton, and they refused to be discouraged from doing it.

They would not be hurried. While John Coulter was amassing a superb collection of wild flowers in the valley and in the marshy shallows around Jenny Lake—named after Jenny Leigh, wife of their guide, Beaver Dick Leigh—they waited until the day which seemed right for the climb. Then they started out with pack mules and horses, up through the dense forests of ponderosa pine and silver-bark fir, to make a temporary base camp near timberline, from which they would start the real climb.

Soon after sunrise on July 29, the group started for the summit. At the beginning there were ten of them, including Frank Bradley, the assistant geologist; Mr. Bechler, the topographer; John Merle Coulter, botanist; James Stevenson; and N. P. Langford, but not all of them reached the top. It proved to be a more difficult climb than it had been feared. A number of them agreed that Jim Bridger had indeed been right.

As they started up the high canyon, ice-water cascades on their left tumbled nearly three hundred feet from the heights of the canyon wall. The men scaled heaps of loose debris and piles of rock where the little brown conies sat and yipped at them with shrill, accusing voices. The rocks seemed alive with the conies, but the wary little animals were difficult to see, yet always

watched from their stern landscape. They were like so many miniature bunnies with large, round, close-held ears; the conies constantly uttered that defiant, warning bark as long as the men were
near. An old, hoary marmot watched blearily from a rockslide.

The men passed the rockslides, reached the first ledge, came
upon the fresh tracks of mountain sheep. The top of the cliff was
a broad valley leading from the canyon wall itself up to the peak,

Alp-lily

a splendid slope of great challenge, bounded on either side by the
rock walls. As they climbed, breathing grew harder. They began
to get into patches of snow. At the edges of the drifts, in the melt-
watered, thin soil, there grew lovely masses of white marsh marigolds and the golden snow buttercups, some standing in water,
some still up to their chins in old snow. It may have been mid-
summer in the valleys below, but it was still spring up on the peak.

Coulter meanwhile was finding small alpine plants, the pink-
flowered cushions of the moss campion, the white Androsace, the
small yellow alpine saxifrage, the tiny, dotted saxifrage, the blue
Milla, and the miniature alp-lily, Lloydia. Then, leaving the slopes
and pools, the men climbed higher into a charming arctic meadow
which was carpeted with the most beautiful of the high-country
flowers—with mats of brilliant blue forget-me-nots on half-inch
stems, with white saxifrage and purple phlox, with small pink clovers and yellow mustards, and the alpine twin-pod which Dr. John
Richardson had found far north in the Arctic.

Evidently picking up insects, violet-green swallows swooped
and curveted over the snowdrifts, while the snow itself was pocked
with innumerable small cavities, in each of which lay a grass-
hopper. They were not dead, only numbed with cold. The air over

the peaks the day before, the men recalled, had been glittering with flying grasshoppers which had seemed to be blown in a cloud above the mountains. Felled by some chill blast, or by the cold alpine night, they had all dropped to the snow and by their own darkness had melted a little way in. Langford, digging out several and finding them only sluggish, felt them begin to stir in the warmth of his hand. As the day's own warmth grew, the whole snowfield was hopping with grasshoppers.

Alpine Sandwort

The men left the alpine meadows and ascended to a chillier level where the little flowers were more meager, now virtually stemless and held against the skull of the mountain. Only the stony, gritty peak itself showed through the sparse growth of low mosses, lichens, and the incredible little tundra plants.

The walls of the valley up which the men had climbed now drew closer together. There was only a gap of fifty or sixty yards to reach a sharp slope of rocky rubbish. They found themselves instead on a narrow crest overlooking another immense canyon which separated the three higher peaks from the mass of the mountains west and north of them. Descent of the steep declivity was necessary before it was possible to climb up the next rise of the final lift of the peak. The trip down the slope was hard. Mr. Bechler turned his ankle and fell with a severe sprain which slowed him down after that. The others found an easier way down. A huge snowdrift, hard-packed and icy, eighty to a hundred feet high, reaching up from below almost to the level where they stood, invited the men to slide. They simply sat down and coasted merrily to the bottom. And above them, almost as if they had come no distance at all, still rose the ever-steeper peak itself. After having climbed so far, they had as yet only reached its base, and this lay beyond a huge lateral spur of rock. It was extremely hard going now. Physical effort in a rarefied atmosphere made every exertion twice as taxing, and the wind was a terrible punishment. It lashed them incessantly. There were pools of ice, and far down in a lone canyon, perhaps never seen before by man, lay a lifeless lake of silent, green-white ice, as remote and chill as a vision on a lunar landscape.

By now, some of the men in desperation had turned back. They had had enough. Bechler himself knew he could climb no farther. His ankle had swelled severely inside his boot and it would be all he could do to get back alive to the camp. He turned off here and, resting frequently, painfully explored a small canyon lower down with Coulter, who had decided to remain where the tundra flowers were.

The five remaining men continued the climb—Bradley, Spencer, Langford, Stevenson, and Hamp. By noon they had reached 11,400 feet. It was harder now, much hindered by steep slopes of snow and ice; some slopes appeared to be composed of hailstones a third of an inch in diameter. A fierce west wind blowing forty to fifty miles an hour swept across the saddle of the peak with such force that the loudest shouts could not be heard a few yards to windward. Stevenson, gasping, realized the danger they all were in. If they did not stay together, they were lost; if one man slipped and needed help, he could never make himself heard or his need known.

Next, Bradley lingered behind. He was expecting West, one of the men who had turned back some time earlier, to bring up a barometer which had been forgotten. West, however, apparently did not enjoy the thought of the climb and he never came with the instrument. Studying the few small alpine plants still in evidence and gathering them for Coulter—one was a new, small Composite —Bradley waited on the slope at 12,000 feet. The other men went on. Spencer and Hamp gave out about three hundred feet below the crest. Langford and Stevenson, filled with a glorious surge of discovery and elation which overrode the worst punishment the wind could bestow, or the altitude and rarefied atmosphere could threaten to life and strength, climbed on.

They scrambled, slipping, stumbling in the loose, shaly rock debris covering the steep slope. There were no more plants here. There was nothing but the snow and rock and debris and wind and sky—and triumph. Stevenson and Langford stood on the top of the Grand Teton. The aneroid barometer read 13,400 feet. (The true height, however, is now placed at 13,747 feet.) First of any men, the two had climbed the Grand Teton.

But were they the first? On the crest, Langford and Stevenson discovered strange slabs of rock which were unlike any kind of material on the upper mountain. The slabs evidently had been arranged in a rude and ancient wall or breastwork. A long time ago, surely a *very* long time ago, someone had carried these rocks to the top of the Grand Teton. Langford and Stevenson, resting on the strange wall, wondered about the mysterious people who had managed to drag rocks to the top of a mountain peak, and wondered even more why they should have done so. Old Jim Bridger had been wrong when he had insisted that no Indian would work so hard to climb a mountain that didn't offer something to eat on the top.

The view from the crest, in spite of the furious wind, was magnificent. The barren, rocky, snowy slopes moved down to the alpine meadows, then to the slopes covered densely with their uniform growth of spruce and fir and the pines, and at last down to the misty valley itself, and to the tiny blue mirrors which were the valley lakes. Against the punishment of wind, the two men could say little, but both of them, almost at the same moment, spoke hoarsely:

"Let's name this peak Mount Hayden. He's climbed . . . more mountains, explored more of the West . . . than any man alive . . . too modest . . . can't help himself this time!"

They shook hands on it, if a trifle numbly, and then, bracing themselves, started down the steep and dangerous slopes which were all the more dangerous on the descent because of the rock-slides and precipices which unexpectedly presented themselves. Reaching Hamp and Spencer who were quite blue with cold, they stumbled down to pick up Bradley, the injured Bechler, and Coulter, with his full collecting case. With great weariness and difficulty they all reached the temporary camp at last. The next day they returned to the valley. Although they might never want to climb the mountain again, they had proved that it could be done.

Frederick Vandeveer Hayden, a lean, slight, reflective man with a close-trimmed beard, had the special ability, perhaps brought about by his own enthusiasm and love for the Western mountains, to attract to himself men of like caliber and a like enthusiasm. By means of the work of his geologists, paleontologists, botanists, and zoologists, and by his own superb dedication and direction, he managed not only to preserve the Yellowstone and the Grand Tetons as national parks, but was the first to make known their complex wildlife.

Among those who got their start on the survey with Hayden in the Yellowstone and Teton country was young John Merle Coulter, who had become so deeply fascinated with the western flora that he could never get enough of it. He was an energetic, restless man who needed to be out in the field collecting, or else working up his collections and writing about them, before he immediately started out for more. He was an excellent botanist for Hayden to have chosen for the 1872 expedition.

Allen and Porter on previous surveys into the Yellowstone had done well with the limited time at their command, but Coulter, with greater opportunity for more detailed work, did even better. Instead of collecting plants as specimens only, he studied the ecology of the area. He soon found that in the Yellowstone-Teton country, plants were divided into three main areas—the plains of Utah and the Teton Basin, the Teton Range and the mountains along the Yellowstone River, and the flora of the geyser basins. No American botanist had ever had the opportunity to identify plants growing in such a uniquely varied place of ecological extremes.

Plants of the Utah and Wyoming plains were simple enough. This was mainly desert or barren land composed largely of mesquite and sagebrush, greasewood, salicornia, white sage, chamisso, and a few cacti. On the plains sloping up to the foothills, plants became more interesting and varied. Here were mariposa lilies, phloxes and gilias, many cacti, the sulphur-yellow Erigonum, Mentzelia, evening primroses, locoweeds in pink and purple, phacelias, and, along the watercourses, some of the splendidly beautiful purple and yellow monkey flowers, several kinds of wild roses, buttercups, Mertensias, gentians, and marsh marigolds, together with masses of the beautiful pink and white Cleome or bee-plant.

Plants of the mountain ranges were sub-alpine and alpine, similar to what Coulter had found on most other high mountains of comparable altitude, yet adding some species or lacking others. He had found the ponderosa pines climbing all the way up to timberline, where they had stopped abruptly. He had found silver-barked fir predominating in the upper reaches, yet with very few firs or spruces remaining as wind-twisted trees at the edge of the tundra.

Flora of the Fire Hole Basin of the Yellowstone also was distinct. It was unique. The hot springs and geysers produced an unnatural soil of geyserite, and at the same time created an artificial warmth. Plants, as if in a hotbed, here grew much ranker than elsewhere. Certain gentians were characterized by having curiously black stems, while the leaves of the little Amarella gentians were unusually dark, the petals a glistening navy-blue which turned black when dried. Plants growing in the geyserite, except for the gentians, were chiefly the Composites—goldenrod, groundsel, Artemisia, pussy toes, yarrow. In some of the hot springs he had found algae, but these neither dried nor traveled well, and could not be taken away for identification under a microscope. Some orange-colored confervoid algae, he decided, no doubt had caused the colors in the geyserite, a fact which is now accepted.

Hayden and the government surveys had opened a new world of amazing natural history to American botanists and zoologists. Ever since Lewis and Clark had crossed to the Pacific, and Townsend and Nuttall had followed; since Vancouver and Menzies, Cook and Nelson, Douglas and Scouler, and Drummond, Richardson, and Franklin had all been exploring and collecting westward and northward, the flowers, trees, and lower plants of Western America had been discovered. One by one named and catalogued, year by year, expedition by expedition, plant by plant, the flora of North America was being found. The government surveys, using a number of men, encompassed exploration over a tremendous space of a tremendous wilderness continent. Trying to learn as much as possible in the shortest time, they worked fast and hard. When the reports went back to the East, other naturalists had no greater dream than to go likewise to the West to find these fascinating new plants and birds in a land which was still so wonderfully new, so refreshingly wild, and still so much in need of detailed exploration and more discovery.

Fern Fossil

13. The Fern in the Rock

It scarcely mattered at the moment that his thumb was bleeding profusely, or that the blazing Illinois summer sunshine reflected violently from the dry shales of the mine dump. He had just given a smart blow with his rock hammer to a nodule of stone, and the nodule, splitting, had fallen open neatly in halves. In each part lay the imprint of a fern of great beauty and immense antiquity.

For the first time since it had dropped into the mud of a Carboniferous swamp, had been encased in an iron-impregnated clay which had turned to stone, this fern-frond in his hand was now exposed to the sunshine of a world it never knew. His were the first human eyes to have ever seen this leaf, for there had been no people when it lived and last knew daylight. Although this was an impression in stone—an intaglio in one half of the nodule, the cameo of it in the other—the actual leaf-fabric seemed to have been carbonized with its original texture, its details, as if outlined in white ink, emphasized with a whitish secretion along each vein and scallop. The fern was an exquisite memento of the past. Two hundred and fifty million years of evolution, change, and time lay encompassed between the opened nodule with the fern inside and the bleeding hand of Leo Lesquereux which held it. He, the fern in the rock, and the free and sunny skies of America were combined in that moment as a result of a remarkable series of events. Commenting, he wrote:

I live among wonders of such an admirable vegetation that nothing, not even a journey along the Rhine, would be comparable to it in splendor (For a fossil botanist)!!!

The "admirable vegetation," found in coal mines and strip mine dumps, was composed of remnants of a glorious flora which had covered great stretches of America during the Carboniferous Era. In mucky shallows following the retreat of the Silurian seas, swamps had fostered a remarkable botany—huge tree ferns with spreading fronds as broad as date palms, giant club mosses towering more than a hundred feet high, great horsetails three feet in diameter at the base, standing erect or leaning of their own weight and size. Like strange Pisa towers patterned with rhythmic corrugations, their tremendous trunks were encircled at each node or joint with a crown of dark bracts, like the iron feathers on a steamboat chimney. The whole structure, topped by strange, gigantic, conelike fruiting bodies or strobili, was fluted and endued with an architectural majesty. In this dream world of a far past there had been beautiful, dense thickets of Sphenophyllum spires, some four feet tall, some less, like flexible stalks strung as with innumerable four-leaf clovers or rosettes. The grotesquely glorious swamps had been inhabited by zooming dragonflies with a thirty-inch wingspread, cockroaches the size of rats scurrying over fallen logs or up tree trunks, and by certain primitive reptiles, complex crustaceans, and elaborate arthropods. There had been no birds in these trees, no mammals in these waters, no warm-blooded animals at all in this strange, warm, prehistoric jungle world.

Lesquereux could visualize it almost as if he had been there. Since he knew how peat formed today, he could thus also conjecture how plant-filled swamps millions of years ago might, by the very bulk of the fallen organic matter decaying in warm mud and water, eventually form coal. Pressure, internal heat, and geological changes all served to alter the deposits and compress them into coal layers far down in the earth.

Not all the plants which became coal lost their identity. Now and again, for a long, long time, there had been certain plant remains found in coal mines. Investigating reports of fern imprints, occasional fossilized roots, stumps, and other plant parts in the depths of American mines, he had gone down with the miners and had searched while they dug. At other times he awaited what the mine owners kindly saved out for him. Sometimes he had rummaged like some oversized woodchuck on the dump piles for this strange, neglected treasure.

When he came to the rich Mazon Creek region in northern Illinois to collect fern fossils with his friend Joseph Even, he saw that a different process had taken place in order to so perfectly preserve the plants and plant fragments which were abundant here. It was through the remarkable specimens sent to him by Even, who lived not far away and could collect there often through the years, that Lesquereux discovered some of the great diversity and perfection in the Mazon Creek fossil flora.

Through examination, conjecture, and a final assurance that he was right, he saw how, in these certain coal swamps and occasionally elsewhere, plant parts had fallen separately into the mud where an ironstone clay deposit in the water and mud evidently had surrounded and encased them. Time, in hardening the clay, had created these unique nodules or concretions. Most of them were shaped like an oval bun and were from two or three to about eighteen inches long, and somewhat flattened. Others were small and round, no bigger than a hickory nut. Although the ironstone composing the concretions was very hard and, when struck broadside with a hammer, usually shattered very irregularly and revealed very little inside, it would, if held on edge and struck along a seamlike area, split neatly in halves. A smart whack with the hammer—and the nodule revealed its secret interior. In almost every one of them lay a fossil leaf, stem, or piece of bark. Occasionally, after an energetic hammering, the rock broke and whatever had been inside was no longer there, having disintegrated at some time in the past two hundred and fifty million years.

He could never get over his astonishment and delight at what lay inside those incredible rocks. It was always a surprise, for there was never any clue as to what each specimen would be. An additional amazement and awe filled him when he realized how very similar many of these ancient and long-extinct fern species were to those which were growing in the Illinois woods not far away.

In plant fossils of all kinds, one always found that strange likeness to familiar, living species. In the beginning, when men in Europe at last could not avoid looking at plant fossils, they tried to explain them away as having been present-day plants which had been petrified by certain ossifying juices in the earth. Martin Luther explained that, during the great flood of Biblical times, when everything not saved on the ark had been lost, a mud coating had formed over everything which lay drowned on the bottom of the waters. These objects had then turned to stone and, the heavier they were, the deeper were they to be found in the earth and its rocks when the waters finally receded and the world emerged. When this theory was forced aside, and as more facts became known, the already incredible dates of geologic history were pushed farther and farther back, until the time of the fern fossils was finally placed in the Carboniferous, two hundred and fifty million years ago.

Plant fossils indeed were a puzzlement. They were not easy for the human mind to admit and accept. In 1558, there was perhaps the earliest record of petrified wood when Georgius Bauer Agricola mentioned it, but he went no further, probably because he did not know anything more to say about his find. Seven years later, Kentmann described leaf impressions in mineral deposits which he found around certain warm springs in Europe, but not until 1664 did anyone dare to admit that he had found actual, logical, irrefutable, lifelike plant forms in rocks. In 1699 came the first picturing of fossil ferns. None was given a name. There were no names to be given; it was still difficult for men to admit to themselves and to each other that such things existed—that it was possible to find plant life preserved in rocks or in coal. In these strange manifestations, there was a faintly sinister implication of the underworld.

Even when scientists like Lesquereux were identifying them, it was confusing and highly difficult to be accurate, because one never found a complete plant, as in living matter, but only fragments. Knowing the variability of some of our modern plants, it was indeed a piece of uncertainty to attempt to identify prehistoric leaves, stems, or bark which might, after all, be parts of the same tree or plant, or were instead from different sources. The example became acute in specimens of beautifully preserved imprints of bark of certain Carboniferous trees, many of which had

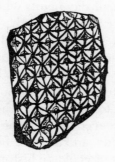

Fossil Bark Imprint
—Lepidodendron obovatum

splendid trunk patterns, particularly the Sigillarias and Lepidodendrons. Yet, because many of these trees changed their designs as they grew older, adding successive layers of both wood and bark, geologists were mistaken into giving different names to each, thinking they were all from separate trees. Not until a complete trunk-section was found fossilized, with each layer present, did they realize to their chagrin that these were in fact all the bark of Lepidodendron, instead of nonexistent species called Knorria, Aspidiopsis, Bergeria, and Aspidaria.

In Germany in 1801, the first real approach to the subject of fossil plants gave birth to a new science called paleobotany. Ernst Friedrich, Baron von Schlotheim, was first to publish a study on the subject, and soon afterward he produced several books on Carboniferous plants. Other Germans pioneered in the exciting new field. In Paris, Adolphe Théodore Brogniart in 1822 was engrossed in a work which, later, would give him the credit as the true founder of paleobotany. Britain joined the new study, and it then came to America when Ebenezer Granger of Zanesville, Ohio, discovered plant fossils in nearby coal deposits. He published an account in the American *Journal of Science and Arts* in 1821.

Thus, in America in the first portion of the nineteenth century, paleobotany was still a new and comparatively little-explored subject. Few men knew anything about it, and those who were pushing their way into the jungle of its complexities were having to be the pioneers and create names, order, and sense in this hitherto unknown and unidentified field. As if he had been working toward it all his life, Leo Lesquereux became a leading spirit in the fascinations and complexities of fossil botany.

There has never been any way in which to determine how a man's inclinations may turn when he is vitally involved in the study of natural history. Sometimes when he is deep in the story of the present-day wildlife, he realizes that there is more to be seen in the past than in life itself, and may turn to the strange story of fossils. These are records of another world, another era, another way of life that ended before man came into existence. Yet fossils, in spite of their obvious antiquity, are inescapably linked with life today by providing a background to the present, a three-dimensional quality to time, and a sense of order, reason, and credibility to the natural world.

Thus, when the doorway to nature swings open, one never knows where the paths of discovery might lead. Besides, it is difficult to be a specialist with a one-track mind which can ignore all else but one subject, for too much is connected in that natural world, one thing with another, in the interlinking relationships whose mysteries have become the study of ecology itself. One never knows . . . but wherever the paths may lead, one has the exhilarating assurance that there is neither an end to the supply of wonder, of the newness lying still undiscovered, or any real end to knowledge.

Leo Lesquereux had been born in 1806 in Neuchâtel, Switzerland, the only child of a manufacturer of watch springs. Since the boy was fond of natural history and books on the subject, he very early had become fascinated with the study of birds and wild flowers. Because of these sentimental leanings, his mother felt that her son would surely study to become a minister, and to this end he was sent to the academy at Neuchâtel. Here, however, Leo had no vocation to the ministry; instead, he discovered an increased yearning toward science. He wanted to go on to the university at Heidelberg, but when his father could not finance this, Leo, not to be thwarted, taught French at Eisenach in Thuringia, Germany, where Bach and Luther had lived, and saved for his ultimate goal.

He was young, brilliant, and amiable; he was handsome, though he often may have wished that his nose were not quite so long. He liked teaching, like his students . . . and he thereupon fell totally in love with one of them. She was the charming and rich Sophia Henrietta. Poor though he was and with no prospects, he married her, and she willingly gave up a social position and money for a life of rich satisfaction with a poor teacher consumed with the excitement of science. She was lovingly confident that he would become one of the best-known scientists of the century.

They went to Neuchâtel, where he had been offered a position at the academy. As he continued to teach, however, he found that something terrible was happening to him. He fought against it, but could do nothing. His hearing was failing. Lesquereux had to strain to hear anything at all, finally had to pretend to understand his students in their recitations because he could no longer make out what they said. His integrity demanded that he resign at once. All his plans, all his hopes and ambitions, were dashed. Young as he was, Leo Lesquereux was rapidly becoming stone-deaf. In desperation, when he could no longer hear his children or his wife, he wrote notes back and forth. He set himself to learn to read lips. As a teacher, however, he was ruined, and teaching was all that he knew how to do.

When he secured a job at manual labor, the brilliant mind which he had cultivated had to lie idle while he used his muscles to make enough money to feed his family. But neither he nor his intellect would quite give up, nor would Sophia Henrietta. Now serving as his interpreter, she learned English and then patiently taught it to him, but because he could hear nothing of how it sounded when uttered, he had a long, slow, frustrating process of learning to read lips when English was spoken. His work and his efforts to adjust occupied his days, but his nights and the weekends were taken over by his brain, by the other part of him which would not be put down by a handicap. Nature and plants were drawing him irresistibly. He said:

I could not stick to that work, and was constantly busy in my hours of rest, that is mostly in the night, with a poor, small microscope, studying mosses, and on Sundays running in the mountains to gather them.

This was engrossing. This was life. Because the mosses were a difficult field which was largely unexplored at that time, they were both a challenge and a fascination. And then, through this interest, as things all his life seemed to turn when matters looked most cruel and hopeless, something turned toward him at last. The government of Neuchâtel was at that time greatly interested in the protection of the peat bogs, for peat was a vital fuel resource. People in need could go out to the bogs and cut it for little or nothing. In relieving the hardships of the poor, this was an important factor to consider.

In working with the mosses, Lesquereux had of course studied the sphagnum in acid bogs left from the Ice Age, understood how slowly it developed, and how it eventually formed layers of peat. He knew something of how the moss kept on growing long after the dead bottom portions were forming the peat itself. Lesquereux made the study, wrote the memoir, and won the prize.

After his paper was published, the great naturalist, Louis Agassiz, saw it, and was at once attracted toward the perception and knowledge displayed in the contents. In his outgoing way, Agassiz sought out the author to compliment him and make his acquaintance. He found the charming Sophia Henrietta, the well-mannered children, and the amiable but stone-deaf Leo Lesquereux.

Agassiz resolved to do something. Through the noted scientist's influence, and because of the prestige of the gold medal awarded to Lesquereux, the King of Prussia soon offered to pay him to undertake a tour of exploration and investigation of the peat bogs in Germany and other European countries. Unutterably delighted in his silent world, Lesquereux took the work and traveled in happy absorption through Germany, Sweden, France, Denmark, Belgium, and Holland—everywhere that peat bogs might be expected to have developed during that series of changes which had taken place at the end of the Ice Age. He returned with a mass of detailed material of great value which he expected to use in a book on the subject.

He was about to set to work on this project—to be financed by the Prussian government—when the great political upheavals of 1848 began. Europe was rent and torn, and nothing was ever to be the same afterward. As the Liberal Party took over, all projects which had been sponsored by the old government were thrown out. The Academy of Neuchâtel where he had taught was also disbanded, and the silent world of Leo Lesquereux fell to pieces once more.

But not all the way, not completely. Sophia Henrietta again held his world together, and he could never despair as long as she was with him. Agassiz had been in America for a year. Hearing of the catastrophes in Europe, he wrote to Lesquereux to join him. There was no other recourse so hopeful to a man with a wife, five children, and no income. He needed to do something at once.

Bound for America, Leo and his family embarked as steerage passengers. After a dreadful voyage they arrived in Boston in September, 1848. Lesquereux, at forty-two, was starting a new life in a new world—a deaf botanist and moss expert who, though he wrote English fluently, had never heard a word of it spoken. With his wife as interpreter and general manager, Lesquereux got his family settled in Boston and eagerly started on the first job which presented itself, one which was totally to his liking. Professor Agassiz, who had just come back from an expedition around Lake Superior, had turned his plant specimens over to Lesquereux to be identified. A room was provided for his use at Harvard University.

This enthralling job completed, he was called to Columbus, Ohio, to work for the wealthy William Starling Sullivant, head of the American bryologists and a noted authority on the mosses. It seemed that every collector returning from an expedition sent his specimens to this arbiter of American moss classification to be studied and identified, and it was more than one man could handle. It was to the advantage of Sullivant, as well as to the tireless plant explorers in the West, that the foremost moss authority in Europe had come to America. Lesquereux moved to Columbus and made his home there for the rest of his life.

Somewhere along the way, among the peat bogs of Europe and over the moss banks of America, Leo Lesquereux discovered plant fossils. He had always studied living plants before; now he became deeply engrossed in a search for plants which had been dead for

Fossil Palm
—Sabal grandifolia

millions of years. He found them in the coal mines of Ohio, plants with beautiful forms and designs of prehistoric beauty, embedded in black anthracite and bituminous layers. Later he found plants in the shales and lignites of the West. Before he was finished, he had helped to reveal the picture of sublime forests in America which had grown, in varying and increasing forms, from the Carboniferous period to the Cretaceous, the Miocene, and the Pliocene, the period just preceding the Ice Age, and had given to America its most glorious flora.

The whole new world of botanical-geological exploration was opening at that time in America, and there were not enough men to do all that needed to be done in revealing the secrets of the past and present. Besides Lesquereux, who became one of the leading authorities on paleobotany, there were Professors Meek and Agassiz, Oswald Heer, and Dr. J. S. Newberry. There was Dr. Frederick Vandeveer Hayden of the Geological Survey, who was fascinated with plants, both alive and fossilized, and who had found a number of new species of Cretaceous fossils in the lignite beds of the Yellowstone region.

Lesquereux, meanwhile, completing his great book entitled *Synopsis of Mosses of the United States,* had ended his work with Sullivant and accepted work with the Geological Survey under David Dale Owen. Since this meant collecting and classifying fossil materials from Illinois, Ohio, and Indiana, it was then that he became well acquainted and enthralled with those surprise packages of marvelous beauty in the Mazon Creek nodules. With

Fossil Sycamore Leaf
—Platanas Haydeni

Joseph Even continually sending him more fine new species, he catalogued the fern fossils. Then the Civil War came on, and when it was over, not only his work with the Owen survey was ended, but it seemed to Lesquereux that just about every other financial resource was also finished.

His sons had been in the jewelry business in Ohio and he was in partnership with them, but the war had wiped out the business. When there really seemed to be no place to turn for a fresh means of earning a living, Frederick Hayden, chief of the United States Geological Survey, hired Leo Lesquereux to study and classify the flora not only of the coal beds, but of the less-known and highly remarkable discoveries of prehistoric plants which were coming to light in the West. Spending several months in the summer of 1872, in company with his son Leo, Jr., Lesquereux worked the lignite beds lying along the Front Range in Colorado. At the same time, Hayden was in Wyoming with his Yellowstone expedition, where he also collected specimens for Lesquereux to study. As more were found, more continually seemed to be coming to light.

Patiently, slowly, expertly, Leo Lesquereux pottered along in the areas which yielded the richest fossil flora in America. Although he was growing old, was totally deaf and was now losing his eyesight, he continued doggedly. He still had the deep, scientific, and emotional thrill in finding glimpses of the long-gone, ancient forests of America which had vanished eons before men. It was his special privilege to reveal them. It was almost like having a secret insight into what had been long ago and would never be seen again. In the leaf specimens which were so beautifully preserved in the lignite, he saw species whose leaves and veins and marred spots remained as if in almost photographic perfection. Leaves from trees were so much like species living today that the likenesses were oddly elusive. It was as if he could scarcely tell where the past left off and the present began, yet the characteristics of the fossil imprints, in spite of their puzzling similarity to living species, were subtly different. It was as if nature had simply been experimenting with different leaf patterns during the Cretaceous and later periods between one hundred and thirty million and a million years ago.

The time of the fern swamps had long since passed. The dicotyledonous plants had come into existence, and nature, as if in an exuberance of creative frenzy to see what might be done with this new and provocative possibility in leaf and flower and fruit, had thus produced numbers of variations on a single theme whose motif might be sycamore or oak or tulip tree or magnolia or sassafras. They were so much like those of today, yet with that difference, so that it was something of an unfinished, hesitant look which they bore. It was as if the designs had not been quite made up then, the patterns not all jelled, as if they were waiting for a decision as to which would remain and which would be discarded. From much of the Cretaceous flora which had not been discarded by a ruthless nature, he began to realize with illumination, our present trees had directly evolved.

At first, only a few species of fossil plants were discovered and given names. Then each explorer and each authority, alerted to the possibilities, found and named more. Lesquereux, in working

for the Geological Survey, was producing one monograph after another in the widely varied field of paleontology. When his flora of
the Carboniferous was finished for the Illinois Geological Survey,
he was at once at work on another which was even greater, a
monograph which dealt with the wonderful flora of the Cretaceous. It would be his finest work.

When he thought that at last it was completed, he had 475
pages of hand-written material entitled *Flora of the Dakota
Group,* illustrated with forty-five quarto plates. Wearily and thankfully, he had sent it all to the director of the Geological Survey.
It was February 21, 1888, and he had described, in all, 350 kinds of
fossil plants which, when he had come to America, had never been
known or even suspected. He was glad he had been given the opportunity to explore these forgotten forests and name their species,
but he was also glad that the task was finished. He was very tired.

Sophia Henrietta had died in 1882. Nothing had been the same
for him after that. Perhaps he had not realized how much he had
depended upon her; perhaps, too, he felt that the very weight of
his own leaning upon her had broken her at last. She was gone, and
for a time he could not sleep, could not work. Hayden feared that
Lesquereux might never complete his study of the great Cretaceous flora.

But the very pressure and the need to finish it were what had
brought him out of his melancholy. Still, his heart was not in it as
it used to be when he could talk things over with her, when he
could get her opinions and approval, when she would read what
he had written, study the plates to be used for the illustrations,
illuminate his being with her delight and her praise.

It was done at last and he was glad. He would, after all, he said,
have to give a thought to making a living . . . but it was to prepare
for dying that the old man spoke.

But although he himself was finished, collectors and their discoveries were not, and they would not let him rest. Before his
manuscript and plates, in the slow procedure of the Government
Printing Office, could be given to the printer, a great new collection of plant fossils came in from Ellsworth County, Kansas. The
Museum of the University of Kansas under Professor F. H. Snow

and Charles H. Sternberg, had been delving into a realm of pure treasure. Now here were thousands of specimens for Professor Lesquereux's classification. And he was so tired. Sadly, he realized that he actually was growing feeble; knew that his days were inevitably coming to an end.

Yet, unable to resist examining a specimen or two, he could not withstand the challenge they offered. They were exciting. They set him on fire. Old Lesquereux, he smiled wryly, was not so far gone after all that he couldn't enjoy a good fossil when it was pre-

Fossil Sassafras
—Sassafras cretaceum

sented to him. He had new life now, new energy. At the age of eighty-two he laid aside his mortal weariness and set to work on the great new task with enthusiasm and the old unquenchable delight.

For the collections were revelatory. They were superb. He could tell very early in his study that there were indeed some fine discoveries here, many of them plants which he had never seen in all his years of exploration in the West. He found himself wishing he were twenty years younger, wishing he could go out again in the field. If Sternberg and Snow could have found all this in Kansas, think how much more must be waiting somewhere else . . . how very much more. If he were only younger.

Well, there was no help for that. This material needed to be added to his monograph, so he asked that the manuscript be sent back to him, that he might make the changes and additions. His children thought he labored too hard, but it was the way he liked to live. Carefully, he worked up all the new matter and added it to his manuscript. He identified every specimen so that it would be unmistakable for future geologists. He supervised the making of beautifully detailed engravings which would fittingly illustrate the work.

When he had finished, he had added 110 new species to the wonderful *Flora of the Dakota Group,* whose presentation of the fossils of the lignites and shales had preserved a picture of forests which had covered so much of America at a time preceding man. The 460 species from that brief area illustrated what might have been the pattern of forests stretching all across America in pre-historic times.

Then, his work finished and the proofs checked, Leo Lesquereux quietly closed his tired eyes. He had played his part in the discovery of America—in the uncovering of an America which no man had ever known before.

14. The Botanist, Thoreau

Before 1857, the wilderness of the Allagash had scarcely known human presence, and much of it remains very little better known today. It was a country of deep forests, bogs, lakes, and of rivers which wandered through the whole, one into the other, in a tremendous, silvery network of waters embroidering a wilderness. It was to this invigorating Maine woods country that, on June 20, 1857, Henry David Thoreau and Edward Hoar headed north from Boston with their plant-collecting cases and presses. They were not so rash as to proceed into the wild country alone. Going by steamer to Bangor, Maine, they immediately contacted a reliable guide among the Penobscots, an Indian who lived at Oldtown with his people.

Joseph Polis, a laconic Penobscot who owned a very nice white clapboard house, was considered to be both a rich man and a very independent one. Depending upon his whim, he was able to take or refuse whatever work was offered. Knowing this, Thoreau doubted that Polis would wish to go with them, and was really very much surprised when the Indian, in an offhand manner, decided he would indeed rather like to go with the two white

239

men into the wilderness so he could shoot a moose for himself. He asked two dollars a day, but Thoreau offered him a dollar and a half, with fifty cents a week for the canoe, and although Thoreau again expected the Indian to refuse the job, the Penobscot agreed. Thoreau, suddenly feeling a delightful lift to the spirits and a lovely anticipation of what was just ahead—as if he were already on the lakes—gave Joseph Polis directions as to where he would find them, and then went with his companion to Bangor to visit a mutual friend until time to leave for the wilderness. The Indian was to come with his canoe on the evening train.

Since Thoreau and Hoar were both serious botanists, this was to be chiefly a plant-hunting expedition. Although, in a romantic period, botany had sometimes become a pleasant, often casual sport, one which was recommended to young ladies as being a gentle interest to occupy them with delicate colors and textures which had nothing rude to offend them, some of the philosophers, the Emersonian men, turned out to be serious and dedicated plant experts. Thoreau might be the first to see the beauty and to measure the meaning in a tree or a flower, but he also wanted to know its scientific name and its correct and unassailable classification. To him, beauty and science must go hand in hand, and he managed therefore to merge philosophy with botany, and scientific exploration with holidays in a very satisfactory blending. Botany might be a recommended pastime for young ladies, but, as two white men and an Indian were about to demonstrate, it could also be a very rugged, masculine occupation.

Thoreau and his friend that afternoon purchased supplies of hard bread, salt pork, coffee, sugar, flour, salt, and a very few other necessities, as well as some India-rubber clothing for wet weather. Thoreau met the seven o'clock train. He really wondered if the redoubtable and independent Joseph Polis might not have changed his mind after all and would fail to appear, but the Indian got off the train at the Bangor depot, saw to unloading his canoe from the baggage car, and nodded to the man who had come to meet him. While Thoreau led the way on foot to his friend's house, three quarters of a mile away, the Indian followed with the canoe over his head.

I tried to enter into conversation with him, but as he was puffing under the weight of the canoe, not having the usual apparatus for carrying it, but, above all, was an Indian, I might as well have been thumping on the bottom of his birch the while. In answer to various observations which I made by way of breaking the ice, he only grunted vaguely from beneath his canoe once or twice so that I knew he was there.

The next morning, in the rain, the three, with the canoe on top of the coach, took the stage as far as the shores of Moosehead

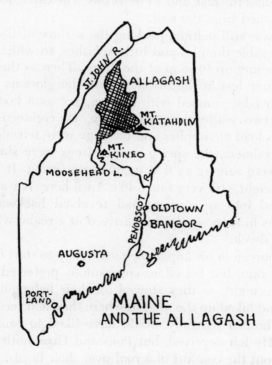

Lake, where they would set off at last to adventure. Thoreau was joyous. He was heading for the wilderness and could hardly wait to get there. The coach trip, however, had been one of frustration. As the carriage rumbled along in the rain, he saw the stalk of a truly magnificent purple-fringed orchis standing nobly beside the road, but, lacking the courage to ask the driver to stop and

let him collect it for his herbarium, he decided against the fool-
hardiness of such a suggestion. This formidable driver, his friend
had warned him, had the reputation for stopping at nothing except to take on a paying passenger. Not even a bear had been
known to halt that stage, much less a mere passenger's request to
pluck a flower. All that rainy day, therefore, Thoreau in botanical
anguish was compelled to see flowers from the windows of the
coach without being able to get one of them. It was most irksome.
Finally, in relief, at half-past eight in the evening, he and his companions reached the lake and were let out. The canoe and baggage
were unstrapped from the roof.

The rain was still pelting, pocking the surface of the gray lake
with innumerable dimples and little splashes, to which the fishes
added by coming up to snap at the drops. Then as the rumble of
the coach grew less in the distance, and the glorious stillness of
the northern lake, rimmed with spruces, lay as a boon on their
spirits, the two white men, an Indian, an eighteen-foot birch
canoe, and a load of supplies and baggage were left alone. In the
cool, wet freshness, the spring peeper frogs were shrilly piping
and toads were singing as if it again were spring. It really was,
Thoreau thought, still very much like April here. It was as if they
had returned into springtime, had revolved backward two or
three months in time, or else had arrived at a realm where spring
perpetually dwelt.

Henry Thoreau in his impatience was ready to start out at once.
He enjoyed rain, but his effete companions preferred something
drier for the night, so they stowed all their belongings beneath
the canoe and hiked up the road to where they had recalled seeing
a tavern. Thoreau wistfully would have liked to camp out that
first night. He felt deprived, but Polis and Hoar both had strong
opinions about the comfort of a roof over their heads. When they
could not help themselves, there would be time enough for camping in the rain. Joe Polis, perhaps thinking of his comfortable
white house at Oldtown, may have shuddered slightly at what he
was about to endure.

By four o'clock the next morning, in the cloudy twilight of a
northern dawn, the rain had ceased. The three left the tavern

without stopping for breakfast; they hurried up the road and launched the canoe in the quiet, pearly, dawn-lit waters of Moosehead Lake. The canoe was packed fore and aft with small baggage, the heaviest stuff in the middle. The craft was so crowded it was difficult for the men to stretch their legs or reach what was needed, but Thoreau saw to it that on top, handy, lay two collecting cases, or vascula. Plant presses took up some more space in the baggage. The presses and cases puzzled the Indian. He shook his head. He could not see the use for them.

As they moved along the shore, the Indian paddling, a few mergansers and spotted sandpipers were startled from the overhanging, root-tangled, rocky shores. When a loon called pensively in the distance, Thoreau knew with certainty that he was in the North. *Medawisla*, the Indian commented. The wild voice of Medawisla, the loon, evoked a deep response in Thoreau. He was feeling an immense pleasure, was relaxing in every muscle and nerve as the soothing dip of the paddle sent the canoe quietly along. He was suddenly, he said in his journal, naturalized, presented with the freedom of the lakes and woods. He was home.

The Indian, after a while, pointedly remarked to no one in particular that a man could not be expected to work very long without eating. Taking the hint, the party, all suddenly very hungry, pulled to shore and made camp, placing the breakfast table strategically near a lovely patch of lavender monkey flowers. Specimens were collected even before the coffee boiled. From a table of freshly peeled birch bark, they ate hard bread, fried salt pork, and strong coffee with plenty of brown sugar but no milk.

The mirror of Moosehead Lake with the looming mass of Mount Kineo reflecting in it were only part of the long journey ahead. Every mile was new and beautiful, and each provided, if not always some new botanical specimens or birds, then some new philosophy for Thoreau, or a new Indian word to be dragged from and identified by the taciturn Penobscot. Thoreau was amassing a growing list of Indian vocabulary. In a day when few naturalists specialized, all things were interesting, all things were grist to their mills. A botanist might easily turn ornithologist at sight of a bird, or a mineralogist in the presence of rocks, or an astronomer if the night was clear, or an ethnologist when an American Indian happened to be around. Thoreau was all of these, and more.

That night when they camped on the shore of the wild and lovely lake, a small tent was their shelter in a light rain. The fire kept them dry, and the smoke held the mosquitoes at bay. The two white men sat by the fire while the Indian gave them some songs in his own tongue. Thoreau gave up trying to write them down and simply listened to this voice from the past.

The camp had been made a little distance in from the lake, with a dark, dense, damp, close forest of spruce and fir around them. Except for the light of the fire, the surroundings seemed as black, endless, and primal as all eternity. As the others slept, Thoreau lay wakeful, pondering the quality of aloneness, knowing that inner clarifying alertness of one to whom the messages of the woods were constantly coming in clear and meaningful tones. He slept at last, but, very late, he awoke suddenly. Something had wakened him. Perhaps, he thought, it had been the loon. A loon could always call up an inner response in him. Or an owl—the owls, too, had their special message. Whatever it was, something pleasant had been hailing him.

Seeing that the fire had fallen apart, he got up to push the brands together and to add some wood, when he discovered, at the edge of the smoldering sticks, a perfect, elliptical ring of white light about five inches in diameter and from an eighth of an inch to a quarter-inch wide. This magic circle was as brilliant as the fire, but was not red like the coals; instead, it burned as with a pure, shining white, like a glowworm's light. Gingerly, he stooped and touched his finger to the ring of fire. It was cold. It had no heat. Straightening up in wonder, he realized that he had found a piece of phosphorescent wood. When Thoreau pared back the bark, he found the interior all aglow with its mysterious light. He cut off some chips with his knife and held them, still glowing with their pale fire, in his hand. In delight he went to waken his companions to see the wonderful thing which he had found.

Even so early in his vacation, Thoreau, by discovery of this lovely surprise, and simply by having seen something he had never seen before, felt repaid for the entire journey.

It could hardly have thrilled me more if it had taken the form of letters, or of the human face. . . . I little thought that there was such a light shining in the darkness of the wilderness for me. . . . I did not regret my not having seen this before, since I now saw it under circumstances so favorable. I was in just the frame of mind to see something wonderful, and this was a phenomenon adequate to my circumstances and expectation, and it put me on the alert to see more like it. . . . I let science slide, and rejoiced in that light as if it had been a fellow-creature. I saw that it was excellent and was very glad to know it was so cheap. A scientific *explanation*, as it is called, would have been altogether out of place there. That is for pale daylight. Science and its retorts would have put me to sleep; it was the opportunity to be ignorant that I improved. It suggested that there was something to be seen if one had eyes. It made a believer of me more than before. I believed that the woods were not tenantless, but choke-full of honest spirits as good as myself any day,—not an empty chamber, in which chemistry was left to work alone, but an inhabited house,—and for a few moments I enjoyed fellowship with them.

In the morning the trio, heading for a spot some thirteen miles away where they would portage to the Penobscot River, set off up the lake. The wind had risen and the lake was rough, so that the canoe was nearly swamped several times before they got it to the shore and beached it and its cargo. It was time for dinner, then, in the rain which was commencing once again. The portage itself was not too long, but since there was so much stuff to transfer, they had to go over it twice, and the afternoon was half over before they reached the river and were reloaded and launched. They paddled downstream on waters which were brimming over the banks from the recent rains.

They had not gone far before Thoreau cried out, pointing to a dark, wooded shore where a stalk of the yellow Canada lily, like a lighted candelabrum whose golden flames were a dozen yellow bells, stood poised in magnificence against the dark background of spruces. It stood six feet high and held its flowers in two great whorls. It seemed a pity to pick it, but when he saw many more standing down the shore, he gladly collected as much of the stalk and blossoms as he could get into his case. The Indian looked at the flower and commented dispassionately that his people used the bulbs of that plant for thickening soups. At this, Thoreau, always ready to try something new, dug several of the white bulbs and tasted one raw—it was rather like the flavor of raw green corn in the ear—and carried others along to use in the next batch of soup which they might decide to make.

Wood Lily

The days went so fast—he could not see how time in the leisurely world of the wilderness could pass so speedily. He scarcely had the time to collect plants and then to arrange them in his presses at night before another day had gone and it was time to sleep again. Sleep was a waste of time in such a precious place.

He wrote plaintively:

Though you have nothing to do but see the country, there's rarely any time to spare, hardly enough to examine a plant, before the night or drowsiness is upon you.

But the nights went so quickly, too. Very early every morning, the white-throated sparrows, piping their thin, insistent whistles from the blueberry bushes and tangles of spruce, awakened him. He was very fond of the white-throats and their songs. "What a glorious time they must have in that wilderness, far from mankind and election day!"

Into Chesuncook Lake, with Mount Katahdin, its peak concealed in clouds, standing in the distance above all the other mountains. The men crossed the lake and at its northeast corner found the Caucomgomox River, and after passing about a mile from the lake on this river, reached the Umbazookskus. Since their route lay up this river, they camped beside it for the night. It had been a strenuous day and they were tired. Morning came all too soon.

It was a very meadowy stream. With its sedges, cotton grass, blue flags, and clumps of narrow-leaved willows along its shores, it reminded Thoreau of the grassy, broad Concord. Much red osier dogwood grew in thickets and tangles, the fruits now turning whitish. Then a distance away from these marshy places, the river grew narrower and suddenly more swift. There were tamaracks on the bank, and tall black spruces dramatic and gaunt on either side, with now and again—as if waiting to be discovered—a splendid stalk of the purple-fringed orchis. The men paused to get a spruce sapling for poling through the shallows, and worked their way up the river.

The Indian had been growing quietly but pointedly impatient with the incessant botanical dallying. He wanted to get into the moose country. Now at last on the tamarack-grown shore he discovered moose tracks. They were such enormous things that Thoreau was impressed, thinking of the obvious hugeness of the animal which must have made them. He would like to see one and agreed with Polis that they might move a little faster to get into the moose country. So they came at last into Umbazookskus Lake, then crossed its shallow waters to the portage leading to Mud Pond. The former lake was at the head of the Penobscot River, the latter leading to the Allagash.

It would be, warned the Indian, a difficult carry. This was the wettest portage in the entire state, he added glumly, and since it was a rainy season anyway, they were bound to find it excessively flooded and very miry. Besides, with all that baggage—his cold black eyes looked over the collecting cases and plant presses—they would again have to go over the carry twice in order to transport everything.

The trail lay through spongy, soggy, sphagnum moss; it was a wet and rocky path through a tremendous coniferous forest of arbor vitae, balsam fir, swamp spruce, and tamarack. With the canoe over his head, the Indian went ahead. He had been over this route before and certainly had no wish to linger or look, especially not to collect plants which he could not eat or use. The others, refusing to be hurried, followed in a more leisurely way. Because of their loads, they were extra slow. Although Thoreau was carrying his full pack of sixty pounds, Edward Hoar chose to carry half of his things partway, put them down and then go back for the rest. Soon, so indefinite was the wet path through the spruces, they lost track of the way the Indian had gone. Forks in the meager ditch in the sphagnum and water which passed for a trail were confusing. Logging paths were additionally puzzling in their sudden branching. The trail (if such it could be designated), as Thoreau related with great relish: ". . . led through an arbor-vitae wilderness of the grimmest character. The great fallen and rotting trees had been cut through and rolled aside, and their huge trunks abutted on the path on each side, while others still lay across it two or three feet high. It was impossible for us to discern the Indian's trail in the elastic moss, which, like a thick carpet, covered every rock and fallen tree, as well as the earth."

The mosquitoes, black flies, and *no-see-ums* were a torment. To keep them from biting, the men rubbed on a wash which they had secured from a Bangor pharmacist. The mixture, composed of sweet oil, oil of turpentine, spearmint oil, and camphor was so disagreeable on their skins that both men decided it was hardly worth the brief protection it sometimes offered.

As Thoreau sat on a wet, mossy, mushroomed stump waiting for his friend, three large gray Canada jays flew past to look at him. Back near the lake he could hear the squealing of an osprey whose nest he had noticed when they started the portage. Otherwise it was very still; now and again a small bird called, but the feeling of aloneness and of wilderness was tremendous. Then, as his now weary companion caught up with him again and they slogged on, they both began to feel even more alone and quite lost in this interminable forest-swamp. They were aware that they had heard nothing of the Indian for a very long time, and had no idea as to where Polis was, or the canoe, or Mud Pond, or the headwaters of the Allagash.

They simply kept on. Passing a bed of wild callas still full of white bloom, they found a trail of sorts which vanished into the open water of an actual swamp, and followed it because they thought it must be the right course and really knew of no better way to go.

We sank a foot deep in water and mud at every step, and sometimes up to our knees, and the trail was almost obliterated, being no more than that a musquash leaves in similar places, when he parts the floating sedge. In fact, it probably was a musquash trail in some places. We concluded that if Mud Pond was as muddy as the approach to it was wet, it certainly deserved the name. It was amusing to behold the dogged and deliberate pace at which we entered that swamp, without interchanging a word, as if determined to go through it, though it should come up to our necks. Having penetrated a considerable distance into this, and found a tussock on which we could deposit our loads, though there was no place to sit, my companion went back for the rest of his pack.

Again Thoreau waited, and as he waited, he botanized. It was a lovely spot for bog plants. On sphagnum hummocks the pitcher plants bloomed; Labrador tea grew around the bog edges, with the remnants of the lavender-pink blossoms of bog laurel and Andromeda still lovely, crisp, and fresh. There also was something new to him. Thoreau had never seen the low birch before (Betula pumila), and he was gratified to find it there. It was not a tree, only a shrub with small, rounded, scalloped leaves, speckled stems, and an arctic look. He thought it would be appropriate to name this place Low Birch Swamp.

He was pondering this pleasant discovery when Hoar came back and with him was the Indian, who was very much worried and quietly reproachful. He could not, complained Joe Polis, really understand how the white men had missed the turning, for he himself had laid a cut branch, very plain to see, to mark the way. *Anyone* could have followed his track in the wet, he added, and he was plaintively unable to see how they had possibly become lost. They were now on the trail to Chamberlain Lake, he added, not Mud Pond, so they would have to come back with him. He obviously had a low opinion of their woodcraft. Thoreau, who liked to pride himself on his own woods-lore and ability to get about in the wilderness, was briefly mortified and more than a little abashed at the Indian's scorn.

Nevertheless, it was very good to be led at last to higher ground, and at the same time to find the great round-leaved orchis. Wet and tired as he was, he had to bend down and measure the width of the green, water-lily-like leaves as they lay back on the moss. Some were nine inches both in diameter and in length, and between them rose a stalk of green-white flowers that stood two feet high.

The dry land was only a brief respite. Again they took to the swampy carry, and the walking, if possible, grew even worse than it had been before. An abundance of fallen timber lay across what passed for a trail. It was a painful route for tired men.

The fallen trees were so numerous, that for long distances the route was through a succession of small yards, where we climbed over fences as high as our heads, down into water often up to our knees, and then over another fence into a second yard, and so on; and going back for his bag my companion once lost his way and came back without it. In many places the canoe would have run if it had not been for the fallen timber. Again it would be more open, but equally wet, too wet for trees to grow, and no place to sit down.

It was a mossy swamp, which it required the long legs of a moose to traverse, and it is very likely that we scared some of them in our transit, though we saw none. It was ready to echo the growl of a bear, the howl of a wolf, or the scream of a panther; but when you get fairly into the middle of one of these grim forests, you are surprised to find that the largest inhabitants are not at home commonly, but have left only a puny red squirrel to bark at you. Generally speaking, a howling wilderness does not howl; it is the imagination of the traveler that does the howling. I did, however, see one dead porcupine; perhaps he had succumbed to the difficulties of the way. These bristly fellows are a very suitable small fruit of such unkempt wilderness.

They were certainly having a wet and wearying time of it, but to Thoreau it obviously was all a glorious experience. The very grimness of the swamp pleased his sense of wildness. Sometime later, he said:

When I would recreate myself, I seek the darkest wood, the thickest and most interminable and, to the citizen, most dismal swamp. I enter a swamp as a sacred place—a *sanctum sanctorum*. There is the strength, the marrow of Nature. The wild-wood covers the virgin-mould,—and the same soil is good for men and for trees. . . . Give me a wildness whose glance no civilization can endure. . . . Hope and the future for me are not in lawns and cultivated fields, not in towns and cities, but in the impervious and quaking swamps.

Eventually, after what seemed an interminable portage, they came to the shores of Apmoojenegamook Lake. They had evidently missed Mud Pond entirely when they became lost and the Indian decided the other route was closer. With considerable relish, Thoreau wrote:

We continued on through alternate mud and water, to the shore of Apmoojenegamook Lake, which we reached in season for a late supper, instead of dining there, as we had expected, having gone without our dinner. It was at least five miles by the way we had come, and as my companion had gone over most of it three times, he had walked fully a dozen miles, bad as it was. In the winter, when the water is frozen, and the snow is four feet deep, it is no doubt a tolerable path for a footman. As it was, I would not have missed that walk for a good deal. If you want an exact recipe for making such a road, take one part Mud Pond, and dilute it with equal parts of Umbazookskus and Apmoojene-gamook; then send a family of musquash through to locate it, look after the grades and culverts, and finish it to their minds, and let a hurricane follow to do the fencing.

It was, nevertheless, a relief to be on dry and solid land again. But, before they could enjoy the dryness, the two white men, fully clothed, walked out into the lake to get the mud cleaned off themselves and to wash their clothes at the same time. They hung the wet laundry on a pole and put on what dry things they had in their packs, then sat by the fire and ate supper which, simple as it was, tasted like a banquet after that extraordinarily strenuous day.

They did not, however, rest well that night. The *no-see-ums* found them. Thoreau began to feel that he had taken a fever. He burned; he felt all inflamed as he lay in his blanket. Then he discovered that the insufferable little insects with their red-hot needle-bites had gotten inside his bedclothes and had been pricking him for hours. To avoid this, the Indian had simply lit his pipe and blown smoke inside his blanket, then had retired into it and pulled the covers over his head.

Finally on the long-sought Allagash, they paddled easily toward another lake, and on and on, through very wild country. It was, therefore, something of a shock, eventually, to come upon houses set back in clearings a distance from the shore. Thoreau, who had trudged up a little path to a farmhouse to buy some brown sugar, got back in time to cut balsam and arbor vitae for his bed before the storm set in again. He said:

It is remarkable with what pure satisfaction the traveler in these woods will reach his camping ground on the eve of a tempestuous night like this, as if he had got to his inn, and rolling himself in his blanket, stretches himself on his six feet by two bed of dripping fir twigs, with a thin sheet of cotton for a roof, snug as a meadow-mouse in its nest. Invariably our best nights were those when it rained, for then we were not troubled with mosquitoes.

Although they had other portages, none were so difficult nor as exciting as the long, unforgettable one. They eventually crossed from the St. John River to the east branch of the Penobscot, and landed at last in a desolate area which about two years before had been devastated and charred by a forest fire. In the ensuing time, nature had worked to repair the burn and in the process had ornamented it with flowers. At a little distance this rolling, open ground, from which thrust gaunt, blackened snags of trees, was all a vast garden of rosy-lavender and red-purple fireweed, *Epilobium*. The men were wading up to their waists in luxuriant billows of bright plumes, flowers resplendent in pure color, the blossoms humming with insects. Among the long wands of flowers were blueberry bushes with ripening fruit, together with wild red raspberries, bracken ferns, and aspen seedlings, plants which are among the first to come into a recent burn.

During the excursion ashore, the Indian waited with the canoe, for Thoreau and his companion had simply climbed up to look at the flowers, pick berries, and speculate on how long it might take before a forest would stand here again. They did not intend to go far or be away very long. But the two men wandered, separated now, and suddenly, ranging along the extremely hazardous rocky cliffs above the rushing waters of the river, Thoreau realized that he had not seen his friend for some time. He had simply vanished. Thoreau, recognizing the ruggedness of the terrain and knowing that Edward Hoar was nearsighted, worried that he might have missed his footing and fallen over a precipice.

He shouted. He hallooed. There was no answer. Growing more and more alarmed, he went back to look across the masses of fireweed which now glowed with the reflected light of a pink sunset. Returning to the cliffs, he shouted again, but there were cascades somewhere below and his voice did not rise far above their noise. He began to wonder how he might tell Edward's family of what had happened, how to explain the tragedy, and how, when they finally found the poor, mangled body, they might contrive to get it out of this rocky, remote wilderness.

Gloomily filled with foreboding, he went down to find the Indian and alert him as to what might have happened. Joe Polis, who did not appear to be particularly worried, set off in an opposite direction to search, but soon returned saying that he had seen nothing of the missing man. Since night was coming on too rapidly for them to dare to hunt much longer along the brink, the pair both hoped that, if the man were still alive, he would not attempt to continue in the dark.

They made a fire on the shore. The Indian was philosophical. ". . . come morning, then we find 'em. No harm,—he make 'em camp. No bad animals here, no gristly bears . . .—warm night,— he well off as you and I."

This was only moderately consoling, for Thoreau knew that his friend might very well be lying injured on the rocks or dead in a gorge. In any case, Edward Hoar was camping alone and hungry. Thoreau could not rid himself of worry, and for a long time lay sleepless in the firelight. He wrote later: "It was the most wild and desolate region we had camped in, where, if anywhere, one might expect to meet with befitting inhabitants, but I heard only the squeak of a nighthawk flitting over. The moon in her first quarter, in the fore part of the night, setting over the bare rocky hills garnished with tall, charred, and hollow stumps or shells of trees, served to reveal the desolation."

Early the next morning, Thoreau urged the Indian up so they might resume the search. The Indian grumbled about the unkindness of sending him off before his breakfast, but Thoreau rather sharply reminded Polis that their friend had had neither supper nor breakfast. They had not gone more than a quarter of a mile when they saw smoke; waving and shouting at them, the missing man stood on a rocky point above the river. He had stuck up the remnants of a lumberjack's shirt on a pole to attract attention, and, although he was very hungry, he was apparently well and unhurt. He had, moreover, in his wandering come upon the very first white fireweed he had ever seen, and had collected this rarity in the burnt lands. While Thoreau examined the find, the Indian, making no audible comment, but only muttering some pointed Penobscot remarks to himself, headed back to the camp.

Joe Polis finally achieved the mission for which he had originally come. Canoeing up a lake, the trio had come upon a cow moose standing placidly among the fallen driftwood and rosy-flowered beds of water smartweed. Flapping her big ears and whisking her tail against the deer flies, she looked comfortable and sleepy. The Indian, suddenly as excited as a young buck, getting himself within range, fired and reloaded and fired again, shooting so wildly that Thoreau could not see how he might possibly have hit anything but the trees and rocks all around the big brown target. But the moose suddenly stumbled and fell. As the great splash subsided, they saw her lying there in the shallows looking, as Thoreau said with sudden pity and distaste, "unexpectedly large and horse-like."

Thoreau, who had never witnessed the destruction of so large a creature, was unhappy to see the splendid animal die. He would always feel that, as he wrote in his *Maine Woods:* "Every creature is better alive than dead, men and moose and pine trees, and he who understands it aright will rather preserve its life than destroy it."

While the Indian, having no such compunction, happily skinned the moose, the white men botanized and looked about for material to make tea. They preferred to be absent from the camp until the bloody business was over. It was much pleasanter browsing in the woods for tea leaves. The Indian had been telling them about the various kinds of wild drinks to be found in the northern woods, so that every night they had tried a different one. They were now ready to have a new kind, and the Indian, still working with precision and dispatch on the moose, told Thoreau to go out and get some wintergreen leaves. There was much of it all about the camp, for this was a forest whose earth was covered with a beautiful evergreen carpet composed of ground pine and moss, snowberry, wintergreen, twinflower, and pyrola. Some of the pink blossoms of the pipsissewa were in bloom. The berries of the Clintonia, held upright on stiff stems a foot tall above the large

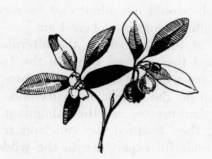

Wintergreen

oval leaves, were turning blue. When Thoreau would have sampled one of the half-ripe fruits, the Indian called sharply to warn him that it was poisonous. In places, the Clintonia was the most abundant plant in the forest.

There was no difficulty in finding the wintergreen's glossy, aromatic, oval leaves. The problem was to tear themselves away from botanizing throughout that charming groundcover for what other things it might provide.

They came back with a bouquet of three-inch wintergreen

stalks. The Indian tied the stems with cedar bark and dropped the bundle to steep into a kettle of boiling water. The resulting tea, well sugared, had a pleasant wintergreen flavor, but Thoreau still thought he liked best the brew made from the snowberry leaves. It had a similar wintergreen flavor, but was more delicate and sweet.

All too soon for Thoreau and Hoar, they came at last down the Penobscot River and to places of civilization. One night as they camped beside the river, they heard an ox sneeze across the water, heard a cock crow before the dawn, felt the presence of men and houses before they were seen. The next day they were passing houses—the town of Lincoln. The Indian, who had eaten too much moose meat after it had begun to spoil in the heat, had been very sick with colic for several days and this had delayed the journey somewhat. Now, approaching familiar country and people again, he got over his stomach-ache and his imminent fear of death, and quickly recovered. They paddled into the lumber-mill country where great booms impounded thousands of logs, and came at last to Oldtown. Joseph Polis would not admit that he was glad to be home. "It makes no difference to me where I am."

They reached his door at four o'clock in the afternoon on August 3, 1857, and stopped there for an hour. In the Indian's neat house, presided over by a quiet Mrs. Polis, the two white men shaved and cleaned up for the benefit of civilization. Then, with their packs, their full plant presses, and their abundant memories of the wild country, they boarded the cars and reached Bangor that night. The wonderful experience in the wilderness was done.

15. John Muir and the Giants

His bed was in the cemetery. The mockingbirds and the chats, the vireos and the titmice had all grown so accustomed to his presence that they no longer scolded at him. The trees, he felt, knew him too—the great live oaks with long draperies of silver-gray Spanish moss—the long-leaf pines with clusters of enormous needles held against the sky—the magnolias with their leathery, glossy leaves that were green above and a golden-brown velvet below. Private, secret, very comfortable besides, the bed in the cemetery was a necessity at this point of emergency in the life of young John Muir. He had known many emergencies, and nature itself, as usual, offered the solution and effected the rescue.

John Muir had come from Scotland with his parents when he was a small boy, and had lived and worked and grown up on the family farm in central Wisconsin. It had not been at all an easy life. These were pioneering times and boys had had to work like men to clear and cultivate the land, and there was little lenience in his stern father's heart for a boy who wanted to roam the woods. Yet, somehow, he found the time and opportunity for this roaming he craved—along the Wisconsin River, around the shores of little Fountain Lake, in the deep woods and bird-filled marshes—so that his experiences in the wild were the foundation upon which all the rest of his life was based.

As Muir grew up, his busy mind turned not only to nature but also toward inventions. One of these was a remarkable clock which was connected to his bed in such a manner that, at the pre-set hour, the clock clanged, the mechanism tilted the bed, and he was thrown out on the floor, presumably wide awake and ready for another day of work.

He managed, by financing his own way, to attend the University of Wisconsin. From there he planned to go on to study medicine at Ann Arbor, Michigan. But the Civil War disrupted his plans. He waited dutifully for his number to be called in the draft. When it did not come, after all, he had to do something to relieve this period of strain and tension, so he went rambling in Canada to collect plants. His enthusiasm for flowers and trees mounted. The botanist in him superseded the physician and the inventor. He lost interest in a career in medicine. The career in invention languished more slowly.

On his return from Canada, however, he had to make a living, and botany, at that time, did not offer it. He went to work in a factory which made broom handles. Immediately, his impatient mind rebelling against inefficiency, he proceeded to invent a machine which would produce broomsticks faster and more neatly than ever before. Then fire struck and destroyed the business— both the broomsticks and his job were summarily ended. Next he went to Indianapolis to find work. There was work elsewhere but he chose Indianapolis because he wanted to see the Indiana prairies and their flowers, and this city was adjacent to them. Here the work went well until tragedy felled him. John Muir in a factory accident was injured and blinded. For a long time, the fear was that John Muir would never see again.

For three interminable and dreadful months he was shut away in a darkened room. When he yearned desperately for all the lovely things in that outdoor world which he might never see again, his new friends brought wild flowers from the prairies and woods, and read to him to keep his mind occupied. At first he had been able to see the flowers and their leaves only with his sensitive fingers, then, with his returning sight, could gradually make out their beloved forms when the light was very slowly admitted to the room.

John Muir had almost lost his world. Now, given back to him as if by a miracle, his sight was doubly precious to him, so that he resolved that he would never again lose contact with the outdoors. As soon as he was released, he was out in the woods. He gloried in the wind and the sun and the summer flowers on the cloud-patterned, grass-rippled prairie. He was hungry to see as much of the world as he could, wanted to devote his life, not to man-made inventions, as he had thought, but to what Muir called the "inventions of God." All the well-turned broom handles in the world could not equal the splendor and perfection of one tall tree. He would rather be a poor-in-pocket naturalist and a happy one, he decided, than a rich industrialist who was forever unfulfilled and unaware of his world.

His great project was to explore the Amazon, for he longed to emulate Humboldt, Spruce, Banks, and the others who had known what great jungles were like, to see the incomparable Amazon and its forests, and the high glory of the Andes. In order to cut his expenses as much as possible, and at the same time see more of his own country, John Muir set off on a thousand-mile walk to the Gulf of Mexico. From Florida he hoped to take passage on a ship bound for South America.

To a man who loved to walk and to whom both people and wild things were friends, a man who was vitally curious about the plants growing from Wisconsin and Indiana to the shores of the sea, walking a thousand miles to observe them was not a hardship but a privilege. He took along little money. He expected his needs to be few, and he didn't wish to be too tempting a mark for thieves who might think he had something worth the stealing. The rest of his money was left with his brother in Portage, Wisconsin, who, after a certain lapse of time, was to send some of it to him in care of the post office in Savannah, Georgia. Then John Muir took the train to Louisville.

As he set out on foot from there on September 1, 1867, the

world lay before him. His heart was light and so was his baggage—
a rubber bag which carried extra underwear, a comb, brush,
towel, a bit of soap, and three small books: a New Testament,
Robert Burns's *Poems,* and Milton's *Paradise Lost.* He also carried a collecting case for plants, and a plant press made of a pair
of light wooden frames with straps to compress between them
the papers containing plant specimens to be dried.

In Kentucky, the new world opened for him when, for the first
time, he saw the glossy, evergreen leaves of the wild mountain

Cardinal Flower

laurel and rhododendron on the hills. He wished that the bushes
might have been in bloom, but just knowing that the sturdy plants
were there upon the flanks of the Cumberlands was enough. When
he found the scarlet plumes of cardinal flower blazing along the
Cumberland River, he was as excited as if he had found an orchid
beside the Amazon.

Reaching the southeastern part of Kentucky, down in that corner where the mountains are highest, he started up the trail leading to the heights of Cumberland Gap. Climbing the long incline,
he appreciated what Daniel Boone and the pioneers must have
felt when they traveled the Wilderness Road as he was doing.
They stood in spirit with him on the summit of the Gap with
Pinnacle Mountain towering still higher. As he looked back into
Kentucky and across into the green wilderness of North Carolina
and Virginia, he felt very much a part of history. In the woods
beside the road he collected specimens of mountain holly, of
American holly, and the remarkable umbrella magnolia whose

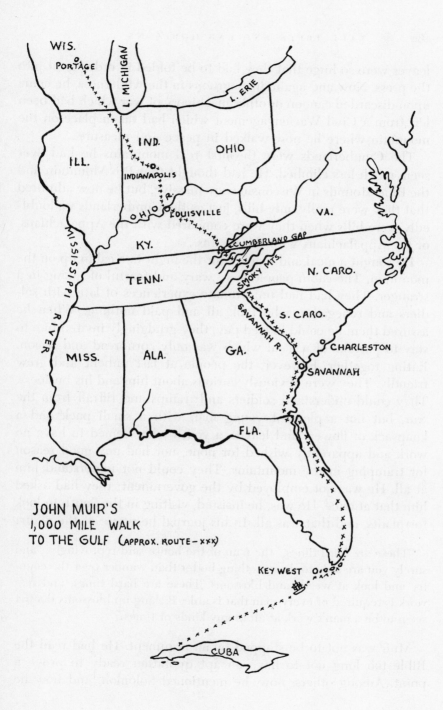

WIS.

PORTAGE

L. MICHIGAN

L. ERIE

IND.

OHIO

ILL.

INDIANAPOLIS

OHIO R.

LOUISVILLE

VA.

MISSISSIPPI RIVER

KY.

CUMBERLAND GAP

SMOKY MTS.

N. CARO.

TENN.

S. CARO.

SAVANNAH R.

MISS.

ALA.

GA.

CHARLESTON

SAVANNAH

FLA.

JOHN MUIR'S
1,000 MILE WALK
TO THE GULF (APPROX. ROUTE- xxx)

KEY WEST

CUBA

262 ✤ TALL TREES AND FAR HORIZONS

leaves were so huge that they had to be folded in order to fit into the press. Now and again, incongruous in the wilderness, he came upon discarded cannon or other machinery of war which had been left from a Civil War engagement which had taken place on the mountain where he now walked in peace and pleasure.

The Cumberlands were the first real mountains he had ever seen, much less climbed. He had thought that Rib Mountain and the Blue Mounds in Wisconsin were massive, but he now admitted that they were really only hills, just as the Cumberlands undoubtedly were hills when they were compared with the Appalachians, or the Appalachians with the Sierras.

He sought a meal and lodging for the night in a cabin up on the mountain. The cabin people were wary and fearful of taking in a stranger. They had had unfortunate experiences of late with soldiers and renegades, who took all and paid nothing. When he assured them he could indeed pay, they grudgingly invited him to stay to partake of a meal which was only cornbread and bacon. Eating together, however, the people at last unbent and grew friendly. They were obviously curious about him and his business. They could understand soldiers and tramps and riffraff from the war, but not a pleasant young man with a small pack and a knapsack of flowers and leaves, a youth who seemed to have no work and apparently wished for none, nor had any good reason for tramping in the mountains. They could not understand him at all. He was not employed by the government; they had asked him that at once. He was, he insisted, visiting in the South to look for plants, and that was all. In his journal he wrote of the affair:

"These are hard times," the man of the house said reprovingly, "and surely you are able to do something better than wander over the country and look at weeds and blossoms. These are hard times, and real work is required of every man that is able. Picking up blossoms doesn't seem to be a man's work at all in any kinds of times."

Muir was not to be stopped by an argument. He had read the Bible too long not to have an apt quotation ready to prove a point. Among others, now, he mentioned Solomon, and how he had openly sanctioned the study of plants; Muir also brought up the matter of the lilies of the field, then concluded:

Now, whose advice am I to take, yours or Christ's? Christ says, "Consider the lilies." You say, "Don't consider them. It isn't worth while for any strong-minded man." The mountain man's tanned face struggled with conflicting thoughts. Then he slowly nodded. He agreed. He thereupon looked with much respect upon John Muir, the botanist, who was on Christ's own work and at His bidding.

One day in the southern mountains, however, he half suspected that his good fortune in finding friendly people was about to change. As he jaunted through a silent, remote valley whose hills brooded closely on either side of the dusty road, he saw ten horsemen riding abreast and quite spanning the road as they came toward him. There was a grim determination in the way they were riding. He felt a strong twinge of alarm, but knowing that he must bluff it out and let the men think he was unafraid, he continued walking toward the riders. The armed, rough-looking, ragged men intently watched him from under their dirty hatbrims and shaggy brows.

With his long, easy strides unfaltering, Muir approached the party, blandly gave them a pleasant "Howdy!", stepped off the road to get around, and was back on the road again, all without pausing or hesitating.

Some distance beyond, not hearing any sounds of his being followed, he risked a quick glance over his shoulder and saw that the ten horsemen had all wheeled about and were still watching as he walked. He decided that the botanic press on his back, from which ferns and leaves protruded untidily, must have given them the idea that he was a poor, itinerant herb doctor out collecting the stuff for his medicines. He, Muir realized thankfully, was obviously not worth robbing.

Five weeks after he had left Louisville, he was in Georgia. Now almost without money, he congratulated himself that he had timed his journey and his expenses so well. On October 8 he went to the Savannah post office to get the funds which his brother had sent him. None had come.

This was a rude jolt. He suddenly felt very poor, very lonely, entirely forgotten. Muir hunted for the cheapest lodging house in which to spend the night. Next day when nothing had come for him, he knew he was going to have to get work, and meanwhile he must eat and sleep as frugally as possible. As a Northerner, however, he did not have an easy day, for he was regarded with dislike and animosity, and was harshly given to understand that southern jobs were for Southerners, not for Yankees. No one would hire him.

Weary and discouraged, much of the glow of the trip gone from him, he found his way along a road margined with live oaks and water oaks, leading out of Savannah, then to open, red-earth fields which once had grown cotton, but which now were neglected and full of jimson weed, mule-tail, and ragweed. He followed the white oyster-shell road for three or four miles and, escaping from the city's unfriendliness, he began to feel a little better. Along the road, and at intervals in the desolate fields, he passed an occasional rickety cabin on broken post foundations, under which dogs and pigs slept. The shattered porch usually held a slat chair or two and a washtub hung on the wall; there was neither glass in the windows nor a door in the opening intended for one. The Negro occupants simply glanced at him as he passed. They were as unfriendly, indifferent, and suspicious as the people in the city.

When he came at last to a beautiful grove of live oaks near the Savannah River, he found within the grove the peace and beauty which seemed so lately to have gone from the South. He had come to Bonaventure Cemetery. An oasis in a land torn and ravished by war and Sherman's march to the sea, it was a lovely thing to approach and a refreshment to enter. Large oaks, swamp holly, purple liatris, and singing birds welcomed him as he crossed the small bridge over a languid brown stream and he found himself within this different world. He felt that any sensible person, having the choice, would prefer to be here with the dead in the lovely graveyard than in the ugliness of the city, with its lazy, disordered way of life, or with the miserable shack people along the dusty, oyster-shell road.

These enormous oaks must have been over a hundred years old. He had never seen trees like them before. They were so different in appearance and habit from the bur oaks and red oaks which he had known in Wisconsin. Live oaks were not particularly tall, only about fifty feet or so, but the trunks were three to five feet in diameter, and the branches were tremendously long and large, spreading horizontally above the avenue and interlocking their twigs and small, oval, evergreen leaves. Each great bough was decorated with a matting of thick mosses, resurrection ferns, tufts of grasses, small air plants, even tiny seedling palmettoes and other trees which had found a rooting in this medium, while the boughs were loaded with long draperies of Spanish moss. It hung in silvery skeins, lavender-shadowed, sun-shot, often reaching eight to ten feet in length, and made up of a maze of small gray tendrils. The skeins of moss came down and mingled with the fan palmettoes which grew around the bases of some of the trees, providing an interesting contrast in texture and form.

Beneath the oaks, the cracked marble tombstones of Bonaventure, each surrounded by an old wrought-iron fence almost hidden under ivy and myrtle, blankets of honeysuckle, and neglected, sprawling roses, occupied only a small part of the protected acreage, while the rest was wild and wooded. At the back still stood the ruins of an old mansion. Beyond lay the sparkle of the river and its broad marshes, where birds foraged in and among the reeds and sedges which filled its shallows. Bald eagles fed on the mud flats or perched on tall pines, while fish crows with their ridiculously high-pitched voices called from the trees or paraded with the willets and eagles on the flats.

Both in the tidal marshes and in the woods, it was all a happy place, a place of singing birds, of many butterflies, of peace. Bonaventure Cemetery was a center of life, not of death. He considered it "Life at work everywhere, obliterating all memory of the confusion of man."

In finding the cemetery, which he decided was the most beautiful spot he had seen in the South, he had also found a temporary solution to his financial problems. After his money had failed to come, and after he had spent precious cash for lodging and a little food, he now had only a dollar and a half left. Not knowing how long it would take before his replenishment came, he could no longer afford a lodging house. He would have to camp out. After the experience he had had with the cockroaches, bedbugs, and dirty bedclothes of his lodgings, camping out would even be preferable.

He considered sleeping on the beach—he had gone out to look at the ocean and had walked in the dunes, bright with goldenrod and soft enough to make a comfortable bed. But when he thought of prowlers, he admitted that he was afraid. Sleeping in the open had been simple enough and safe enough in the mountain wilderness; it was a different matter here. Near the cities of the war-battered South of Reconstruction times, marauders both white and Negro were about and no man was safe alone at night. To have made his bed beside the resounding waters of the Atlantic Ocean would have been a great delight under other circumstances, but not now.

John Muir realized that Bonaventure Cemetery would be the ideal place for a man in his situation. Since he was not at all superstitious, he had no qualms about bunking with the long-gone dead in their peaceful, beautiful haven, while the best part of the plan was that few marauders, if any, would ever think to look for a living victim in a cemetery, nor dare to venture into it at night on any other business.

The sun had already set in an autumn glow which transformed the marshes into pools of apricot and pink and gilt and apple-green. He hurried down the oyster-shell road, now lavender in that magic light, and past the tumble-down shacks which were embellished with their own glow of sunset. No one looked out to see him. A scrawny dog barked halfheartedly. He crossed the little bridge and stepped into the cemetery for the night.

Beneath the oaks it was almost dark. The marble tombstones now and again gleamed out of the purple gloom. In no wind, the moss hung in long, immobile draperies; leaves in the thickets of sparkleberry bushes (one of the huckleberries) glistened like clusters of crystals in the light of the rising moon. It filtered light through cracks in the canopy, glimmered down past the great arms of the oaks, and silvered the tombstones and the huge gray trunks of the trees. He was tired, but he was so enchanted with the scene that he walked about for a little while before finding a comfortable place for his bed. Beneath an oak, a little mound served as a pillow—next morning he discovered that it had been a small grave-mound on which his head had rested all night. He slept fairly well, although now and again he wakened as certain large beetles whose feet felt prickly scrambled across his face.

Yet he slept. When he woke, the moon had long since set and the sun, streaming into the cemetery, was lighting with gold the

Sparkleberry

draperies of moss, illuminating the great oak trunks, and shining on the tombstones. The birds were singing splendidly and, as he stirred, they discovered him. Some hooded warblers and white-eyed vireos were much disturbed and came down close, scolding and fussing about to examine him. He heard the eagles out over the marsh, and the cackling and piping of sandpipers, willets, and plovers on the tidal flats. The laughing gulls flew and dipped, and the handsome royal terns with white wings and bodies and scarlet beaks flashed and dived in the dazzling sunlight.

He could hear people far away, the calls of Negroes somewhere out on the road, the distant sounds of Savannah. He was hungry, but he was so bemused with the beauty of his surroundings that food was unimportant. For a little while he sat there watching the light change, watching the birds and squirrels. He ate a couple of crackers for his breakfast before he stepped unobtrusively out of the cemetery and walked into Savannah to see if his money had come.

It had not.

Well, there was no help for it. He was going to have to accept the fact that he would be sleeping in Bonaventure Cemetery for an indefinite length of time. He decided to go back before dark and choose a place where he could be better concealed and more comfortable. He prepared his nest in a thicket of sparkleberry bushes near the right bank of the Savannah River, not far from where the eagles had their roost and a quantity of smaller birds dwelt.

His house was fun to make. With four adjacent bushes as corner posts, he tied their branches to the middle to form a roof covering a space four or five feet long and about three feet wide. Then he gathered armfuls of rushes from the river marsh and thatched his roof, and pulled down a quantity of moss to spread on the floor for a soft, dry, and fragrant bed. It was really very neat and would have done credit to an Indian or a bear.

As he lay there luxuriantly that night, watching the moonlight through his doorway, he wondered how long a man could live on water and half a dozen crackers a day.

He had slept five nights in his graveyard home when he began to realize that even by spending only three or four cents a day on crackers, his final twenty-five cents was not going to last very much longer. Although he tried again and again to find some work, employment in Savannah was denied to him. His northern accent was against him.

As he went to town on the sixth morning, he was so weak and giddy that he staggered like a drunkard and dimly hoped that no one would notice his peculiar gait and question him. It seemed that the dazzling white oyster-shell road ahead of him was rising and falling in undulations, like waves on the shore, and the streams along the ditches all seemed to be running uphill. He made his way to the post office once more and, when he asked again, now almost without hope, if his money had come, the clerk replied that it had. The clerk seemed unimpressed with this overwhelming event.

The money was indeed here, but how, the clerk was asking with a maddening deliberation and caution, could Muir identify himself as Muir? After all, the post office was obliged to hand over the money only to a certain specific John Muir, and how was anyone to know who that person was unless he had some identification to prove it?

Muir, feeling his knees quivering with weakness and holding them stiff so he would not shake too much, groped in his pocket for his brother's letter. He handed it to the postal clerk, saying that it stated the amount of money to be sent, where it came from, the day it was put into the post office at Portage. He thought that would be enough, but the clerk was still cautious. This rough-looking, shaggy man in clothes which had obviously been slept in for some time, and probably not in any bed, either, and the bearded face, the gaunt looks—no one knew who he was or what his background was.

The clerk parried. He did not, he said, feel that that was enough for identification. The man after all could have stolen the letter, might have murdered the rightful owner. After all, could he himself prove that he was really John Muir?

Muir stiffened. He simply would not collapse before he had proved his identity. Looking at the letter again, he pointed to a sentence—"I hope you are having a good time and finding many new plants." This, obviously, he told the clerk with what he felt was truly monumental patience, certainly indicated that this John Muir must be a botanist. He himself was a botanist. If the clerk wished to question him on any botanical point, he would be glad to answer and thus prove that he, and only he, could be Muir the botanist. A thief who had stolen the letter might pretend to be John Muir, but he could never have pilfered the latter's knowledge of botany.

The clerk gave up. He consulted with a man in the inner office. Muir waited. He held on to the edge of the counter so that he could stay erect with dignity.

When the clerk came forth at last, he brought the package to Muir, who took it thankfully and hurried out. He was so hungry that when he saw a woman selling gingerbread in the street outside the post office, he invested in a chunk immediately, and ate it as he walked. Then he found a place to secure a real meal. All of a sudden, all over again, John Muir felt wonderful. He wrote, "Thus my 'marching through Georgia' terminated handsomely in a jubilee of bread."

Revived and replenished, he vacated his lair in the cemetery and was on his way down into Florida, to the Keys. Here he explored and collected until he fell ill from malaria. Nursed at a plantation, he went on, as soon as he was able, to Cuba for more botanizing.

He had hoped to secure passage on a ship leaving Havana for Brazil, but, when none materialized, he finally gave up for the moment his dream of South America. He sailed instead for New York on a freighter carrying oranges.

New York terrified him. Many years later, thinking back on that time, he wrote: "I can make my exhilarated way over an unknown ice-field or sure-footedly up a titanic gorge, but in the terrible canyons of New York, I am a pitiful, unrelated atom that loses itself at once."

Escape from the frightening canyons lay in a steerage passage costing forty dollars on a ship bound for San Francisco, via the Isthmus of Panama. It was neither a good nor a comfortable voyage and he preferred afterward not to think about it, but the journey across the Isthmus, that narrow neck of mountainous land which the Spanish, English, and others had been crossing in the paths of Balboa, Wafer, and Dampier, was fascinating to a naturalist. Passengers for San Francisco disembarked on the Atlantic side of the Isthmus of Panama and took a small railroad train through the mountains to meet a ship waiting on the Pacific side. This train ride itself was almost as good as exploring the jungles of South America. He would never forget the brilliant tropical flowers and the tremendous trees along the first twenty miles of the Chagres River as the little train ground its way toward the Pacific Coast. He was afraid to leave the observation platform for fear he might miss something.

As soon as he had landed in San Francisco, John Muir asked the way to the nearest wilderness—and headed at once for Yosemite Valley. It could be recommended as the most remote, the wildest, and the most wonderful part of California, and it was as yet

largely unexplored. On the Oakland ferry he met a young Englishman named Chilwell who also wanted to see the Yosemite
country, so they joined company.

The usual way for anyone to venture there at that time was to
go by river steamer to Stockton, take passage on the stage for
Coulterville, then hire horses for the mountains. But John Muir,
having all the time in the world, had forsaken the urgency of
clocks or calendars. As he and his companion in the beautiful
spring weather walked up the lovely Santa Clara Valley to San
Jose, he felt himself losing the last traces of his discontent with
cities and his disappointment in not visiting the Amazon. The
whole valley, he said, was a lake of light flooded with that tran-
scendence of the California sunshine with its special quality of
clarity and brilliance.

Mockingbirds and grosbeaks sang, almond and plum blossoms
perfumed the wind. The air itself possessed a special flavor which
thrilled Muir in a way he had never known before. This country
was where he had been coming all his life, and he had arrived at
last. He had come home.

Chilwell and Muir went up Pacheco Pass into the mountains
to a height of fifteen hundred feet and from this point they looked
from the summit into paradise. In the distance stood the snowy
Sierras, and in the valley of the San Joaquin blossomed masses of
flowers. He had never known flowers to literally paint the landscape as these did. It was, as far as the mountains, all a vast and
level flower garden that lay like a lake of gold where the California poppies and Composites and mustards, accented by blue
lupine, purple penstemon, and pink verbena blossomed by millions. From the eastern rim of this golden garden rose gently
sloping foothills dotted with oaks, to a broad belt of coniferous
forests, with the white peaks above all. The nearest mountain
must have been more than a hundred and fifty miles away, but
the atmosphere was so clear and so deceiving in that clarity that
it seemed to be within easy walking distance.

All the way up into the mountains, Muir was finding flowers
and trees which were new to him. There were such charmers as
mariposa lilies, the Sierra lilac, the baby blue-eyes, and oaks with

small bluish foliage and white bark, oaks which grew separated one from another in parklike areas which were not man-made openings but natural ones. There were oaks whose tiny leaves had sharp spines like holly, and some oaks with long, narrow acorns. He watched the gaudy California woodpeckers hammering holes in trees and posts, then energetically tamping an acorn into each hole, tight as a cork in a jug. Climbing higher, the men came among the ponderosa pines standing beside the clear waters of the Merced. There were lovely azaleas and a flowering dogwood which was different from the one he had known in the East and the Middle West; it was the same California dogwood which Nuttall had found and which had been named for him.

The road ended. When Chilwell cautiously wondered if they should proceed, he found the eager Muir already half a dozen yards ahead of him, and so the pair followed a trail which led to a height of six thousand feet, into drifts of snow and into a remarkable forest of giants—sugar pines, silver firs, Douglas firs, and incense cedars. The snow still lay ten feet deep in places in the forest, snow vastly deep and pure, and, wherever it had melted on the southward slopes, he found flowers.

The pair came upon a deserted cabin and decided to spend the night so that they might look about still further in the vicinity next morning. His companion, letting Muir botanize, cleared away the snow from in front of the door and got it open. He cut some soft, ferny boughs of silver fir and made beds for the two, but Muir would really rather have made his bed outdoors. Chilwell, he commented sadly in his journal, had the house habit.

And so John Muir came into Yosemite and made it his own. He could not get enough of it. He reveled in its size and splendor, in waterfall and flower meadow, in lake and river, in Half Dome and El Capitan, in Sentinel Dome and the Cascade Cliffs, in the Cathedral Rocks and Tenaya Canyon and in the remoteness of the lovely Tuolumne Meadows. When he and Chilwell finally climbed up the trail to Wawona and the grove of Mariposa Sequoias, he had reached his greatest experience with trees. For, if he had thought he had seen large trees before, they had been as nothing compared with those dwelling up here—structures of vast antiquity, enormous as pillars in some vast Olympian temple. He and Chilwell found themselves speaking in whispers; it seemed somehow not proper to talk aloud in the shadowed and awesome presence of the forest.

That night they camped in the grove and made a fire below the great trees so that the red-brown trunks seemed more than ever like illuminated temple columns, the ferns picked out in small fountains of light, the young trees spotlighted, while the stars seemed to have come down closer for having filtered their glitter through the feathered boughs of these immensely tall Sequoias. Two hundred and thirty feet high, they were, and some were thirty-six feet in diameter. Later Muir learned that they were more than four thousand years old, had begun to grow far back in the days when the story of mankind itself was young, and so encompassed man himself in their timeless and enduring structure. Although he would see redwoods and Douglas firs which were even taller, none was so broad in the base, so massive in the

Sequoia Cones

trunk, nor as old as the hoary Sequoias. Crusted here and there with their charteuse-green lichens and bright emerald mosses, they dominated the mountains.

These were his trees; his, Muir's. The godlike giants of Yosemite had become instantly part of his flesh and spirit. It seemed to him that he had been waiting all of his life for this moment and for this vital homecoming.

Simply wandering from tree to tree and looking up, the two men spent long days in the forest. They had long since run out of adjectives. Six weeks passed in roaming through Yosemite and yet they had not seen it all. Chilwell at last went home, but Muir, obtaining work with a sheepherder, stayed on, first as a shearer, then as shepherd when it was time to take the flocks to the upper meadows in the spring. He never enjoyed his woolly charges; he disliked the way in which they grazed everything down to the roots and ruined the mountain meadows and woods, but it was the sheep which made it possible for him to stay, and so he had to tolerate them.

After he had spent a winter in the rigorous storms and snows of the Yosemite country, he emerged more devoted than ever to the wonders of this stupendous place. The snows too had been great to see, their white garmenting of the coniferous trees superb, each tree taking the snow according to its character and form. With the coming of another spring and forsaking his shepherd's job, Muir set about exploring farther, collecting flowers, looking for some of the northern species he had known in Wisconsin and Canada, particularly the charming little twinflower, or Linnaea, and the snowberry, *Gaultheria hispidula*. Nostalgically, he said, "I would rather see these two children of the evergreen woods than all the twenty-seven species of palm which Agassiz met on the Amazon."

Muir set himself to be the protector, the defender, and the jealous owner of the Yosemite. The place itself with its giant Sequoias had first been seen by men of the Joseph Walker Expedition in 1833, but a trapper, Galen Clark, is credited with having really discovered and made them known in 1857. The area had been set aside by Congress in 1864 as a recreational area or state park for California, but it was not a national park, and there was great danger that part of it, together with the adjacent unprotected areas, might be destroyed by private interests which were even then making plans for cutting the Sequoias and sugar pines, damming the rivers, and turning the meadows into farmland.

Many people by then had made the difficult but rewarding pilgrimage up to the mountains to feast their eyes and souls upon the stirring landscape and the massive trees. With the help of these dedicated persons, Muir worked furiously to make it a national park. He knew that Yellowstone had been so designated in 1871. Not until 1890, however, through Muir's great efforts, his endless attacks by letter and article, his pounding the urgent facts into the consciousness and the conscience of Congress, was he successful. Within a week's time, in that momentous year, Yosemite and Sequoia had both been made into national parks, and so had another area of giant trees, General Grant National Park, now part of King's Canyon National Park. If John Muir had no other means of being remembered, his three national parks, as well as Muir Woods, a national monument preserving some of the coast redwoods, would have been more than enough to make him an American immortal. He was a small man who had turned his mental powers and his eloquence, augmenting his profession as a botanist, to becoming the champion and the guardian of the biggest and grandest and some of the oldest trees in North America.

Muir, however, soon had another problem. In advertising the wonders of the big trees and the tremendous terrain of Yosemite, he found that the more he wrote and spoke about it, the more people came into his valley to see it. He was full of contradictions. He wanted the area saved, demanded that people appreciate it, but he also did not like to have too many of them coming into and spoiling his precious wilderness or invading his privacy. Like Daniel Boone, he wanted plenty of elbowroom, plenty of breathing space, plenty of wilderness. Then to his relief, he found that the tourists stayed mostly in the easy valleys; they went to see the Vernal Falls and the Bridal Veil, admired Half Dome and El Capitan, stayed at the Ah-Wah-Nee and Wawona Hotels, perhaps pitched a tent in the camping grounds, and perhaps got up to Tuolumne Meadows, and rode to the Mariposa Grove to see the Sequoias before going away again. This place was so enormous that he found it very easy indeed to avoid the crowds.

"They do not annoy me. I revolve in pathless places and in higher rocks than *the world* and his ribbony wife can reach."

In eight years of study and living in Yosemite, he became deeply interested in another aspect of the place. He was proving his new theory of the origin of the great valley, of the rounded cliffs, of everything that made it unique and impressive. It had been believed to have been created by volcanic upheaval, but he was not at all sure that he now believed this. Muir had discovered living glaciers in remote portions of the park, had studied their present action on their surroundings, then had discovered the same sort of striated bedrock, moraines, outwashes, and horse-shoe-shaped valleys in the park where there were then no living glaciers. This was proof to him that all of Yosemite was indeed the result of ancient glaciers which had carved canyons, gouged valleys, sheared cliffs, and laid the foundations for the plants and animals which inhabit it today.

To get a wider view and prove his point, he made trips into Alaska, went on the government ship *Corwin* * to Wrangell Island in the Arctic Ocean. He became a noted glaciologist whose theories of Yosemite finally were accepted.

To Yosemite and to Muir—often as much for the one as for the other—noted people from many places came to visit. Although he liked a measure of solitude, and in fact required it for a healthy mind and way of life, he also loved company, especially the intellectuals of the day. The people he liked and admired were those who had kindred spirits and like interests. And they were coming—the scientists, the writers, and the naturalists. Muir and Yosemite drew them.

He thought he would fairly burst with excitement when he heard that Ralph Waldo Emerson, whose writings and philosophy he had so much admired and lived by, had left his sanctum in New England and was actually at the hotel in Yosemite. Muir was suddenly too shy to meet him, but Emerson sought him in his sugar-pine cabin near Lower Yosemite Falls. Emerson and Muir became friends at once; they delighted to be together. Exhilarated by the exchange of thoughts and philosophies, they visited every day, and they talked and talked and talked.

When the time drew near for his departure, Emerson invited his new friend to join the party which was going out by way of the Mariposa Sequoias. Muir would go, he bargained, looking at the craggy-faced old literary lion, if Mr. Emerson would camp with him that night in the grove. He would, he promised, build a

* See Chapter 18, *Men, Birds and Adventure*, by Virginia S. Eifert.

glorious campfire that would light up the templed boles of the big trees. At the younger man's enthusiasm, Emerson became almost excited. Muir had a very infectious kind of excitement about him. Emerson promised to camp with him beneath the Sequoias.

But Muir, after his brief delight, was disillusioned. Emerson was sixty-eight, Muir only thirty-three. The people in charge of the author's itinerary were being very careful of him. He was precious, must be guarded, and he thus had hardly any time to express a wish for himself or do anything on his own. Too old to be assertive, he simply was obedient to their wishes.

Thus, when the party had reached the Wawona Hotel, Muir was surprised when everyone dismounted. They had not yet come to the place under the trees where they were to camp. When Muir asked what had happened to change the plans, he was greeted coldly by the committee, who said in reproving tones that it would never do to let Mr. Emerson lie out on the ground all night. If Mr. Emerson should take cold—

Muir argued fiercely that no one ever took cold up in those pure forests. Only in towns did they get sick. There wasn't a single cough or sneeze in the whole of the Sierras. He argued that he would build a great fire which would permit no one to feel chilly, spoke of the fragrance of flames of the Sequoia wood itself, and how transfigured the big trees would be by firelight. These people would not have really known the Sequoias until they had seen them in the light of a campfire, and they should not deprive the great man of this privilege. But these, too, he saw sadly, had the house habit in the worst way, and they were actually afraid of the woods. He felt it was an unhappy commentary on Boston culture and the glorious transcendentalism of Emerson and his kind.

The next day, however, Emerson and his party came to the Mariposa Grove and walked about for an hour or so like so many tourists, admiring the big trees but scarcely comprehending them, he knew. Then the people were on their way, while Muir, deeply disappointed and somehow empty, saw his hero leave. He felt inexpressibly hurt and deprived. He had been so sure that Emerson, of all men, would have been the quickest to see the real mountains and the higher meaning of the trees, then would go back and write about them as only he was able to do. Muir stood and watched the guests depart. As dusk came on, he built himself a fire and sat alone beside it.

"And though lonesome for the first time in these forests, I quickly took heart again—the trees had not gone to Boston, nor the birds, and as I sat by the fire, Emerson was still with me in spirit, though I never again saw him in the flesh."

Muir could not always stay in the mountains. He was obliged to go down to Oakland in the lowlands to write a book. He went on journeys. He wrote more books. He got married and acquired a ranch.

But Yosemite always drew him back, and he was ready to drop everything he was doing and go there whenever anyone came from the East and needed a personally conducted tour. He relished the botanists and naturalists who came. In resounding conversations which still must be echoing around the red-brown boles of the Sequoias, they sat around John Muir's campfires and talked.

But it was an all-round remarkable occasion in 1877 when Asa Gray, the leading American botanist, and Sir Joseph Dalton Hooker, president of the Royal Horticultural Society of London and son of Sir William Jackson Hooker, visited Yosemite and Muir. As a botanical explorer, Sir Joseph had traveled widely. He was one of the first to develop and begin to prove the theory of plant and animal relationships which became the science of ecology.

He had come to America to visit Asa Gray and see some of the American West. Frederick Vandeveer Hayden was financing the expedition west for the illustrious men, partially to aid his own ends in securing the assistance of two of the most noted botanists of the day. It was Dr. Gray's first trip to the far West; he had classified countless numbers of specimens of western plants sent to him by men of the expeditions, but he had never seen most of these species in their native haunts, or in any other form than dried plants. He was sixty-seven, and it was time he did so.

The two botanists collected furiously. Before the party left Colorado for Utah, Hooker's plant presses already contained more than five hundred species.

In California, Yosemite and John Muir were high on the itinerary. Muir, totally delighted to have such knowledgeable guests, showed them everything, and then took them on an expedition to Mount Shasta to collect plants. One night when they had made camp in a flowery meadow surrounded by majestic silver firs—Muir liked especially well to camp near silver firs because of the way in which the firelight illuminated the tiers of branches and the feathery needles—he wrote:

After supper I built a big fire, and the flowers and the trees, wondrously illumined, seemed to come forward and look on and listen as we talked. Gray told many a story of his life and work on the Atlantic Alleghenies and in Harvard University; and Hooker told of his travels in the Himalayas, and of his work with Tyndall and Huxley and grand old Darwin. And of course, we talked of trees, argued the relationship of different species, etc. I remember that Sir Joseph who in his long active life had traveled through all the great forests of the world, admitted, in reply to a question of mine, that in grandeur, variety, and beauty, no forest on the globe rivaled the great coniferous forests of my much-loved Sierra. But it was not what was said in praise of our majestic Sequoias and Cedars, Firs, and Pines, that was memorable that night. No; it was what was said of that lowly, fragrant namesake of Linnaeus—*Linnaea borealis*.

After a pause in the flow of our botanic conversation that great night, the like of which was never to be enjoyed by us again (for we soon separated and Gray died), as if speaking suddenly out of another country, Gray said, "Muir, why have you not found Linnaea in California? It must be here or hereabouts on the northern boundary of the Sierra."

In reply, I said I had not forgotten Linnaea—That fragrant little plant, making carpets beneath the cool woods of Canada around the Great Lakes, has been a favorite of mine ever since I began to wander.

"Well, nevertheless," said Gray, "the blessed fellow must be living hereabouts no great distance off."

Next morning Gray continued his work on the Shasta flanks, while Hooker and I made an excursion to the westward over one of the icy-cold branches of the river, paved with cobblestones; and after we had forded it we noticed a great carpet on the bank, made of something we did not at first recognize, for it was not in bloom. Hooker, bestowing a keen botanic look on it, said, "what is that?" then stooped and plucked a specimen of it, and said, "Isn't that Linnaea? It's awfully like it." Then finding some of the withered flowers, he exclaimed, "It's Linnaea!"

This was the first time the blessed plant was recognized within the bounds of California, and it would seem that Gray had felt its presence the night before, on the mountain ten miles away.

Other men came to see him. John Torrey was one; so was C. C. Parry, discoverer of many western plants. John Burroughs came to California and let his old friend Muir show him the glories of his Yosemite. Muir had visited Slabsides years before, on his last occasion to be in Massachusetts, and had written frequently to encourage Burroughs to come out to the West and see what it had to offer. And so at last the Sage of Slabsides, also known as John o' Birds, came out to visit John o' the Mountains.

It was May when the two walked together in Muir's great valley and looked at the blue-purple shadows on the mountains, and the sunlight striking rainbows in the waterfalls, and listened to the finches and ouzels and grosbeaks singing, and stood at the feet of the giant Sequoia trees. They examined the evidences of glaciers, roamed awhile in the upper meadows, saw the bears and the deer and a moose or two. They had a thoroughly happy and strenuous time—two elderly gentlemen in flowing beards who had the unquenchable enthusiasm for nature in their hearts. Burroughs was seventy-two, Muir a year younger.

Burroughs wrote afterward: ". . . one could live in Yosemite and find life sweet. It is like a great house in which one could find a nook where he could make his nest, looked down upon by the gods of the granite ages."

This was Muir's own nest, as closely his, though so much larger, as the nest he had had beneath the sparkleberry bushes in Bonaventure Cemetery outside Savannah. From Yosemite, as from the haven under the live oaks, he could look out at the world and find a clearer understanding of it and all its wild things. In the Sierras, as well as around little Fountain Lake in Wisconsin, or in the mountains of Kentucky, or among the live oaks of Georgia, Muir had found some of the secret, timeless wonder of tall trees in a green and vital America.

*Flower Bud and Leaves
of Cucumber Magnolia*

16. Tall Trees and Far Horizons

It was spring. A certain subtle fragrance on the wind told of poplar catkins blossoming, of young leaves coming out on the willows, and of elms in bloom. It spoke of earth and old leaves which had been well watered by last night's April rain, then warmed by the morning sun. Without a doubt, spring had truly come.

There was no need for the rollicking, just-back-from-the-South medleys of the first brown thrasher in the top of the pear tree. There was no need for the wave of robin song which had rolled like a tide over the town at half-past four in the morning, nor even for the purple martins cutting great sweeps across the sky and chortling in downright mirth and good feeling.

It was really spring—the dandelions were in bloom on the lawn, wild plum thickets were full of bees and pollen, and the woods were perfectly carpeted with one of the most exciting visions of the year. There they were, the manifestations of April. Spring beauties, tenuous and frail, their thin petals veined with pink, had surmounted winter to come into bloom. Bloodroot, its crinkled, gray-green leaves folded like small, veiny hands around

281

the stems, blossomed like small white water lilies above the floor of the woods. Hepaticas, some lavender, some pink, some white, stood above the furry nubbins of new leaves, while trout lilies formed beds of miniature white and yellow Easter lilies. There were blue violets, white violets, yellow violets, with their perennial violet charm. Dutchman's breeches stood above feathery tufts of leaves. Lavender phlox scented the air. The mayapple umbrellas unfurled, and the whole woodland was full of blossoms. With music and motion and sunshine, with all the fulfillment of spring, April had redecorated the landscape.

The identity of anything in the woods really was not needful to the enjoyment of the sight and sound and smell of spring, nor for a feeling of delight in the aspect of the living woods. One might stand among trees and flowers and simply appreciate them for what they were, the embodiment of springtime in the North, where all things were taking place in an orderly sequence. The flowers were materializing out of winter by a process evolved by natural forces when the Ice Age had created extremely short summers and very long winters. In order for flowers to form seeds and for the leaves to make food for storing in the roots, to trigger the coming cycle of flowers, plants had had to work fast, had had to make two seasons do the work of one. But these facts were unimportant to appreciate the miracles of spring. There was no real need for anyone to know the name of any flower, or any bird, or why it was there, or why certain things were not there, or anything. All that was necessary on an exuberant April morning was emotion, a feeling, an awareness, and a thankfulness for being alive to see it happening all over again. The inevitability of April surges into the senses on a morning like this. It gathers a person up as part of the broad-scale arrangement of time, life, and the seasons.

Fortunately for the study of natural history and the development of botany, zoology, and related knowledge, there always have been people who couldn't just stand there. Appreciation or not, they had to do something about it. They were the ones who were filled with the driving urge to find out the names of everything or, if there weren't any names to be found, it was their mission to see that some were created. Plant by plant, bird by bird,

creature by creature, for more than four hundred American spring-
times, the knowledge of American natural history grew. Across
this continent there were thousands of things which had to be
found, learned, and named. Because of this, it is possible today
to know something about the flowers and trees and birds, and all
the other wild things on a springtime morning, or at any other
moment in the year. Back of all the excellent guidebooks lie those
several centuries of discovery and hard work by men who were
the true discoverers of America itself.

With all of these energetic precedents, it is very hard for most
people to just stand there. Because of this restless curiosity to
know about things, the story of the past and all the long-expended
energies of the botanists have become part of our own lives and
experiences, have made it possible for all men to have a fuller
appreciation of each springtime and of each facet of the good
green year.

Except in a very few wonderful and jealously guarded areas,
our once untraveled and unknown wilderness may be largely
gone; yet, there is a portion of wilderness and some of the mystery
of the unknown in every woods, every marsh, every meadow and
bog, in every three-feet-square section of prairie land, on every
hill or shore. Thoreau said, "to be awake is to be alive," and to be
awake and alive to these immediate natural surroundings is to be
insatiably curious about them and immensely fulfilled by learning
their stories. That curiosity provides a key to a particularly pleasant and rewarding life of unending mental and physical adventure.
It can begin in childhood, and it frequently continues as a perpetual enrichment far into old age.

Botany may be, perhaps, all things to all people. It might be a
source of only casual interest to provide a pleasant hobby. It can
also very easily result in an absorbing lifetime of study, collecting,
and travel in search of new plants. An awareness of plants may
simply provide something to look at on the way to and from work
every day—the personalities and identities of trees along the city
street, flowers in gardens, small plants in the cracks of the side-
walk, all yield a certain treasure. An interest in plants also means
going to the woods to catch the first lovely burst of spring blos-

soming, with always the chance that, even in the best-known woodland, there will be, this year, something new. A different sort of interest in plants means a careful collecting program throughout the growing season to make a comprehensive study of a specific area, or in a particular group of plants. It can be the hobby of taking specimens of tree buds in winter, perhaps making sketches of them, perhaps photographing them in highly enlarged form, perhaps putting the twigs in water and watching the intimate procedure as the bud scales expand and miniature new leaves emerge.

Since plants have the agreeable habit of staying in one place, with little of the flitting and vanishing of the more restless forms of life, they are ideal to be learned about on the spot. Plants provide a challenging field of photography—in black and white, in color slides, in color prints, or in time-lapse photography. The plant photographer seldom needs either a blind or a telephoto lens, or quite as much of an infinite patience in waiting as photographers of the more lively forms of wildlife usually do.

There is, however, still plenty of challenge, because flower photography is not as easy as it looks. Even though plants do not get up and run about, they can nevertheless become extraordinarily animated just as one is focusing for a really fine close-up shot.

Just as there are in every science and hobby, there are specialists in the field of botany. One may choose to study the wild orchids as young Oakes Ames did; or the Composites, or the violets, or dozens of other fascinating families. One might choose what could be termed a post-graduate course in plants, accompanied by a considerable amount of mental cudgeling, uncertainty, and ultimate satisfaction, by collecting and identifying the complex sedges, grasses, mosses, lichens, or haws, as men like Schweinitz, Muhlenbergh, or Lesquereux did. Or one may range the desert with Engelmann and Wislizenus and Bigelow to look for cacti; or take up such fascinating things as fungi, ferns, slime molds, water plants, parasites, saprophytes, bacteria, tropical plants, house plants, or garden plants, or go about with a hand lens to discover the astonishing details in one's back yard or in the nearest woods.

This interest in what grows is what gives incentive to many a vacation and motor trip, to the walk in the woods or the climb up the mountain. It is what gives identity and meaning to every sort of landscape, to all that tremendous green, flowery heritage bequeathed to us by all the botanists who did the spade work, who located and gave names—the neat, comfortable, Latin binomials—to all the thousands of plants in America.

To the inheritors of that great store of knowledge and inspiration which has been the result of adventure, pain, hunger, danger, death, and the ultimate triumph of truth among the men who, from Hariot to Muir and the botanists of today, discovered and studied American plants, there remains the same source of personal delight, excitement, and fulfillment. As long as trees grow and flowers bloom, this remains one of the natural assets of America and the living world.

It is only needful to watch springtime come dramatically to the northern parts of America after a long winter, or, more gently to see it appear, in its own perfumed way, to the South, to know all over again the thrills of the other plant hunters from the beginning of man's observation of his world. It is of very little matter if thousands of plants have been discovered and named in America, if one has seen only a few of them. In everyone lies the potential of his own rewarding personal discoveries which, after all, are what the earlier naturalists sought and knew.

The fact that someone else was first to find the Calypso orchid or the red maple tree or the wild camas does not tarnish the inner pleasure of seeing one's own first red maple in its autumn scarlet, finding pale blue camas beds along a prairie road, or the emotional impact of coming upon a secretive little orchid growing in some dark, cold cedar wood. These are moments of pure discovery.

David Douglas was no doubt as much inwardly illuminated at finding the fir which Menzies had discovered, or that Meriwether Lewis had rediscovered, as if he himself had been the very first man to behold it. The discovery was not so much of the tree, but in seeing it himself. It is surely this same illumination and enrichment which comes in any experience of finding something which one has never seen before.

Those plant discoverers have provided the excellent flower and tree guides, the moss guides, mushroom books, the manuals of algae and aquatics, of prehistoric vegetation, of grasses, and of ferns. They are profusely illustrated and the plants are described so that, unlike the predicament of John Bartram and Jane Colden, a student today can understand the language. In those botany books, indelibly placed in inconspicuous lettering, are the names of the men and women who were the plant explorers and plant discoverers of America. Their names are found in the Latin binomials or are represented simply as an initial or an abbreviation just following the Latin name of a plant. The discoverers have thus autographed their finds.

Whenever we roam the nearest woodland or park, walk from home to town and look into gardens along the way, or when we vacation in the northern woods, explore the Olympic rain forest, travel beside the shores of Maine, walk on the crest of Fall River Pass, follow the Mississippi, go into the Everglades, ascend the Sierras, or traverse the plains and deserts, we know that these people have been there before us. In their hardships, their pioneering, and in their triumphs, they have smoothed the path by giving names and meaning to everything we are apt to see.

We share with them a kinship as with those of like interests anywhere, whether it is with friends of today or with people who may have lived a century or two ago. They enjoyed the same things we now enjoy, and they wrote about these things in terms which stay forever green and alive. They are our comfortable, personal association with men long gone and with wildernesses long tamed, whose tall trees are our trees, and whose far horizons are always beyond us still, waiting to be explored.

Bloodroot

Bibliography

1. Land of the Sassafras Tree

Fenton, William N., *Contacts Between Iroquois Herbalism and Modern Medicine*. Bureau of American Ethnology. Annual Report of the Smithsonian Institution, Washington, 1941. p. 503.

Josselyn, John, *New-Englands Rarities Discovered, in Birds, Beasts, Fishes, Serpents, and Plants of the Country*, with introductory notes by Edward Tuckerman. William Veazie, 1865.

Lorant, Stefan, *The New World*. Duell, Sloan and Pearce, 1946.

Peattie, Donald Culross, *A Natural History of Trees of Eastern and Central North America*. Houghton Mifflin Company, Boston, 1950.

Sargent, Charles Sprague, *Manual of the Trees of North America*. Houghton Mifflin Company, Boston, 1905. (Also in two volumes, by Dover Publications, Inc., New York, 1961, paperbound.)

Tyler, Lyon Gardiner, ed., *Narratives of Early Virginia*, Vol. 9, 1606-1625. Charles Scribner's Sons, New York. 1907.

2. Linnaeus—a Name for Naming

Arber, Agnes, *Herbals, Their Origin and Evolution, 1470-1670*. Cambridge. 1938.

Bailey, Liberty Hyde, *How Plants Get Their Names*. Macmillan Company. 1933.

Dorrance, Anne, *Green Cargoes*. Doubleday, Doran and Co., Inc., Garden City, New York. 1945.

Gleason, Henry A., *The New Britton and Brown Illustrated Flora of the*

Northeastern United States and Adjacent Canada. New York Botanical Garden. 1952.

Hawks, Ellison, *Pioneers of Plant Study.* Macmillan Company. 1928.

Linné, Carl von, *Lachesis Lapponica,* Lapland Observations, from Warner Library, Volume 15. Toronto: Glasgow. 1917.

Peattie, Donald Culross, *Green Laurels.* Houghton Mifflin Company, Boston. 1950.

3. John Bartram, the King's Botanist

Cruickshank, Helen Gere, *John and William Bartram's America.* Devin-Adair Company, New York. 1957.

Eifert, Virginia S., *Men, Birds and Adventure.* Dodd, Mead & Company, New York. 1962. (In this book see also chapters on Catesby, Nuttall, Du Pratz, Lewis and Clark, John Muir, and the Railroad Expeditions.)

Faris, John T., *The Romance of Forgotten Men.* Harper and Brothers, New York. 1928.

Goodnight, Clarence and Marie, "Alexander Garden, Physician and Naturalist." *Nature Magazine,* December, 1947, Vol. 40, No. 10, p. 525.

Sutton, Ann and Myron, *Exploring with the Bartrams.* Rand McNally Company, Chicago. 1963.

4. Jane Colden, First Woman Botanist

Catesby, Mark, *Natural History of Carolina,* 2 Volumes. London, 1771.

Colden, Jane, *Botanic Manuscript,* selections publ. by the Garden Club of Orange and Dutchess Counties, New York. Edited by H. W. Rickett. Chanticleer Press, New York. 1963.

Frick, G. F., *Mark Catesby.* University of Illinois Press. 1961.

5. Peter Kalm and the Mountain Laurel

Benson, Adolph, *Peter Kalm's Travels,* 2 Volumes. Wilson-Erickson Inc., New York. 1937.

Dorrance, Anne, *Green Cargoes.* Doubleday, Doran and Co., Inc., Garden City, New York. 1945.

6. Michaux the Botanist and the Spanish Conspiracy

Fox, Frances Margaret, *Flowers and Their Travels.* Bobbs-Merrill Company, Indianapolis. 1936.

Minnegerode, Meade, *Jefferson, Friend of France.* G. P. Putnam's Sons. 1928.

Michaux, André, *Travels of André Michaux. Early Western Travels,* Volume 3, R. G. Thwaites. Arthur H. Clark Company. 1905.

7. Nuttall

Bradbury, John, *Travels. Early Western Travels,* Volume 5, R. G. Thwaites. Arthur H. Clark Company. 1905.

Nuttall, Thomas, *Journals of Travels into the Arkansas Territory, 1819. Early Western Travels,* R. G. Thwaites. Arthur H. Clark Company. 1905.

Nuttall, Thomas, *The North American Sylva*, 3 Volumes. Robert Smith, Philadelphia. 1854.

Townsend, John Kirk, *Narrative of a Journey Across the Rocky Mountains, 1839. Early Western Travels*, R. G. Thwaites. Arthur H. Clark Company. 1905.

8. Mystery of the Mississippi

Du Pratz, Antoine Simon Le Page, *History of Louisiana, of the Western Parts of Virginia and Carolina*. London, 1774. Annotated edition, The Pelican Press, Inc., New Orleans.

Eifert, Virginia S., *Louis Jolliet, Explorer of Rivers*. Dodd, Mead & Company, New York. 1961.

Eifert, Virginia S., *Mississippi Calling*. Dodd, Mead & Company, New York. 1957.

Eifert, Virginia S., *River World: Wildlife of the Mississippi*. Dodd, Mead & Company, New York. 1959.

Pike, Zebulon, *Expedition of Zebulon Pike;* edited by Elliott Coues, 2 Volumes. Francis P. Harper, New York. 1895.

Schoolcraft, Henry Rowe, *Expedition to Lake Itasca; the Discovery of the Source of the Mississippi River, 1832.* Michigan State University Press. 1958.

Schoolcraft, Henry Rowe, *Narrative Journal of Travels Through the Northwestern Regions of the United States to the Sources of the Mississippi River, in the year 1820.* Michigan State College, 1953.

9. David Douglas

Armstrong, Margaret, *Field Book of Western Wild Flowers*. G. P. Putnam's Sons, New York. 1915.

Cook, James, *Voyages Around the World*, 7 Volumes. London. 1809.

Douglas, David, *Journal Kept by David Douglas during his Travels in North America, 1823-1827.* London. 1914.

Franklin, Sir John, *Narrative of a Second Arctic Expedition to the Shores of the Polar Sea, in the years 1825, 1826, 1827.* London. 1828.

Franklin, Sir John, *Thirty Years in the Arctic Regions*. George Cooper, Publisher, New York. 1859.

Godwin, George, *Vancouver: A Life, 1757-1798.* Appleton, New York. 1931.

Harvey, Athelstan George, *Douglas of the Fir*. Harvard University Press. 1947.

Higman, H. W., "The Man of Grass," *Nature Magazine,* March, 1949, Vol. 42, No. 3, p. 138.

Higman, H. W., "First Flower," *Nature Magazine,* January, 1947, Vol. 40, No. 1, p. 20.

Smith, Edward, *The Life of Sir Joseph Banks*. John Lane, London. 1909.

10. The Plant Men

Duboc, Jessie L., "Montana's Historic Bitterroot," *Nature Magazine,* October, 1947, Vol. 40, No. 8, p. 426.

Dupree, A. Hunter, *Asa Gray, 1810-1888.* Harvard University Press. 1959.

Fox, Frances Margaret, *Flowers and Their Travels.* Bobbs-Merrill Company, Indianapolis. 1936.

Gleason, Henry A., *The New Britton and Brown Illustrated Flora of the Northeastern United States and Adjacent Canada,* 3 Volumes. New York Botanical Garden, 1952.

Gray, Asa, *Manual of Botany, A Handbook of the Flowering Plants and Ferns of the Central and Northeastern United States and Adjacent Canada.* Eighth edition; largely rewritten and expanded by Merritt Lyndon Fernald, 1950. American Book Company, New York-San Francisco. 1950.

Lewis, Meriwether, and Clark, William, *Original Journals of the Lewis and Clark Expedition, 1803-06.* 7 Volumes. Antiquarian Press, New York. 1959.

Long, Stephen Harriman, *Expedition to the Rocky Mountains. Early Western Travels,* R. G. Thwaites. Arthur H. Clark Company. 1905.

Reinsmith, Winston H., "Torreya, Tree with Possibilities," *Nature Magazine,* March, 1945, Vol. 38, No. 3, p. 153.

Rodger, Andrew Denny, *American Botany, 1873-1892, Decades of Transition.* Princeton University Press. 1944.

Rodger, Andrew Denny, *John Torrey, A Story of North American Botany.* Princeton University Press. 1942.

11. A Route for Rails

Armstrong, Margaret, *Field Book of Western Wild Flowers.* G. P. Putnam's Sons. 1915.

Hafen, Le Roy R., *The Frémont Disaster, 1848-1849.* Arthur H. Clark Company. 1960.

Ives, Joseph C., *Report of the Colorado River of the West, 1858-59.* Washington, D.C. 1859.

Sitgreaves, Lorenzo L., *Report of Exploration Down the Zuñi and Colorado Rivers.* Washington, D.C. 1854.

Whipple, Amiel Weeks, *Report of Explorations for a Railway Route near the Thirty-fifth Parallel of North Latitude from the Mississippi River to the Pacific Ocean: 1853-4. Volume III, Reports of Explorations and Surveys to Ascertain the most practicable and economical route for a railroad, made under the direction of the Secretary of War.* Washington. 1856.

12. The Teton-Yellowstone Wilderness

Hayden, F. V., *Preliminary Report of the U.S. Geological Survey of Montana and Portions of the Adjacent Territory.* Washington. 1872.

Hayden, F. V., *U.S. Geological Survey of the Territories. Report of Explorations for the Year 1872 in Montana, Idaho, Wyoming, and Utah.* Washington. 1873.

McDougall, W. B., and Baggley, Herma A., *Plants of Yellowstone National Park.* U.S. Government Printing Office. Washington. 1936.

Melbo, Irving R., *Our Country's National Parks.* Bobbs-Merrill Company, Indianapolis. 1941.

Nelson, Ruth Ashton, *Plants of Rocky Mountain National Park*. U.S. Department of the Interior, National Park Service. U.S. Government Printing Office. Washington. 1953.

13. The Fern in the Rock

Berry, Edward Wilber, *Tree Ancestors*. Williams and Wilkins Company. Baltimore. 1923.

Hayden, F. V., *Preliminary Report of the U.S. Geological Survey of Montana and Portions of Adjacent Territory*. Washington. 1872.

Janssen, Raymond E., *Leaves and Stems from Fossil Forests*. Popular Science Series, Volume I. Illinois State Museum. 1939. Revised 1947. Paperbound.

Lesquereux, Leo, *Flora of the Dakota Group*. Government Printing Office. Washington. 1891.

Lesquereux, Leo, Geological Survey of Illinois. *Paleobotany: Description of Plants*. Published by authority of the Legislature of the State of Illinois. 1866.

Newberry, J. S., *The Later Extinct Floras of North America*, U.S. Geological Survey Monograph XXV. Government Printing Office. Washington. 1898.

14. The Botanist, Thoreau

Thoreau, Henry David, *The Maine Woods*. Edited by Ellery Channing. 1864.

Thoreau, Henry David, *Walden,* and other writings. Edited by Brooks Atkinson. Random House. 1950.

15. John Muir and the Giants

Badé, William Frederick, *Life and Letters of John Muir*. Houghton Mifflin Company. Boston. 1924.

Fry, Walter, and White, John S., *Big Trees*. Stanford University Press. 1930.

Melbo, Irving R., *Our Country's National Parks*. Bobbs-Merrill Company, Indianapolis. 1941.

Muir, John, *Cruise of the Corwin*. Houghton Mifflin Company, Boston. 1917.

Muir, John, *John of the Mountains:* unpublished journals of John Muir. Houghton Mifflin Company, Boston. 1938.

Teale, Edwin Way, *The Wilderness World of John Muir*. Houghton Mifflin Company, Boston. 1954.

Wiley, Farida A., *John Burroughs' America*. Devin-Adair Company, New York. 1951.

16. Tall Trees and Far Horizons

Brown, William H., *The Plant Kingdom:* Textbook of General Botany. Ginn and Company, Boston. 1935.

Eifert, Virginia S., and Voss, John, *Illinois Wild Flowers*. Popular Science Series, Volume 2. Illinois State Museum, Springfield. 1951.

Gleason, Henry A., *The New Britton and Brown Illustrated Flora of the Northeastern United States and Adjacent Canada*. New York Botanical Garden. 1952.

House, Homer D., *Wild Flowers*. Macmillan Company, New York. 1935.

Mathews, F. Schuyler, *Field Book of American Trees and Shrubs*. G. P. Putnam's Sons, New York. 1915.

Mathews, F. Schuyler, *Field Book of American Wild Flowers*. G. P. Putnam's Sons, New York. Rev. ed., 1950.

Sargent, Charles Sprague, *Manual of the Trees of North America*. Houghton Mifflin Company, Boston. 1905. (Also in two volumes, paperbound, by Dover Publications, Inc., New York, 1961.)

Index